Praise for *Building Solutions wit* *Compact Framework*

"If you've been looking for one book on the .NET Compact Framework that will teach you everything you need to know, look no further—this is it! Jon and Dan do a wonderful job of covering this content, so that readers are sure to find much enlightenment within these pages."

—Derek Ferguson, Chief Technology Evangelist, Expand Beyond Corporation

"While many technical books focus on API details, this book covers architectures and best practices. It highlights the special concerns for .NET mobile development. It is an excellent book for experienced .NET developers who want to transfer their skills to the red-hot mobility arena."

—Michael Yuan, University of Texas at Austin

"Moving to the .NET Compact Framework is about to become much easier with the aid of this book. The samples and supplied utility code will help you get a running start on your own mobile development projects."

—Kent Sharkey, Program Manager, MSDN

"This book is the starting place for development on devices using the Microsoft Compact Framework. Whether you are doing business or commercial applications, this book is the resource to kick-start your journey into mobile development."

—Stephen Forte, CTO, Corzen, Inc.

"The .NET Compact Framework brings the power of .NET programming to mobile devices. Dan and Jon's book is an excellent resource to help build your programming skills in the mobility area."

—Kevin Lisota, Product Manager, Mobile Devices Division, Microsoft Corporation

"Mobile applications are truly a new class of applications and having feature-rich development tools and infrastructures like VS .NET and the .NET Compact Framework is only half the battle of learning to build them. It's great to see Dan and Jon apply their in-depth knowledge of the .NET Compact

Framework to cover the important topics that everybody learning to build mobile applications should understand."

—Craig Neable, .NET Evangelist,
Microsoft Corporation

"Mobile solutions and their development present new challenges for enterprises, especially with disconnected/synchronization scenarios. Leveraging best practices and planning mobile development properly are essential to the successful implementation. *Building Solutions with the Microsoft .NET Compact Framework* provides a solid introduction to mobile development and best practices. It is a great resource for experienced architects and developers who are new to mobile solution development."

—Steve Milroy, Solutions Director, Mobility and Emerging Technologies,
Immedient Corporation

Building Solutions with the Microsoft .NET Compact Framework

Building Solutions with the Microsoft .NET Compact Framework

Architecture and Best Practices for Mobile Development

Dan Fox

Jon Box

Addison-Wesley

Boston • San Francisco • New York • Toronto • Montreal
London • Munich • Paris • Madrid
Capetown • Sydney • Tokyo • Singapore • Mexico City

The publisher offers discounts on this book when ordered in quantity for bulk purchases and special sales. For more information, please contact:

U.S. Corporate and Government Sales
(800) 382-3419
corpsales@pearsontechgroup.com

For sales outside of the U.S., please contact:

International Sales
(317) 581-3793
international@pearsontechgroup.com

Visit Addison-Wesley on the Web: www.awprofessional.com

Library of Congress Cataloging-in-Publication Data

Fox, Dan (Dan L.)
Building solutions with the Microsoft .NET Compact Framework : architecture and best practices for mobile development / Dan Fox, Jon Box.
p. cm.
ISBN 0-321-19788-7 (pbk. : alk. paper)
1. Mobile computing. 2. Microsoft .NET Framework. I. Box, Jon. II. Title.

QA76.59.F555 2003
005.2'768—dc22

2003018536

ISBN 0-321-19788-7
Text printed on recycled paper
1 2 3 4 5 6 7 8 9 10—CRS—0706050403
First printing, October 2003

Pro Laura, homo doctus in
se semper divitias habet

—Dan Fox

For Tonya and sons,
the reason I pursue

—Jon Box

Contents

Foreword

Welcome to the new age of rapid application development (RAD) for mobile devices!

The Visual Studio .NET 2003 and the Microsoft .NET Compact Framework enable a whole new era of smart device application development. The .NET Compact Framework greatly simplifies the process of creating and deploying applications to mobile devices, while also allowing you to take full advantage of the device that you are targeting.

Developers of all types, ranging from hobbyists to enterprise developers, will benefit from these simplified processes and will enjoy a development experience on par with desktop and server application development. Dan and Jon have done a great job of introducing these new mobile development technologies to developers from all backgrounds. Seasoned desktop and server .NET developers and experienced mobile developers alike will benefit from this book.

This book will teach you everything you need to know to build fast, reliable applications for devices.

Welcome to the .NET mobile community!

Jonathan Wells
Technical Product Manager
.NET Compact Framework
Microsoft Corporation
October 2003

Preface

A Brave New World

The last three years have been exciting times for architects and developers. Microsoft, in the spring of 2000, first announced a vision for computing termed Next Generation Windows Services (NGWS), which in February of 2002 materialized as Visual Studio .NET (VS .NET) and the Microsoft Windows .NET Framework. These products brought object-oriented programming to the world of Web development and helped mainstream the paradigm of connected applications using XML Web Services.

However, it could be argued that the biggest boon for developers working with Microsoft development tools such as Visual Basic 6.0 (VB), Visual Interdev, Active Server Pages (ASP), and Visual C++ was that the introduction of the .NET Framework unified the programming model. Now, developers could use a common Integrated Development Environment (IDE), core languages, and tools and reap the benefits of rapid application development (RAD) programming inherent in VB, Web server applications with ASP, and powerful object-oriented development with Visual C++. This unification created opportunities for developers to extend their skills into new application areas by leveraging a core set of class libraries tied to a runtime engine and development environment.

Fast-forward a little more than a year. Microsoft has now built on the foundation laid by v1.0 of the Windows .NET Framework by releasing v1.1 and Visual Studio .NET 2003. In this release not only has Microsoft consolidated features released after the initial release to manufacturing (RTM), including ADO.NET Data Providers for ODBC and Oracle, Visual J#, and ASP.NET Mobile Controls—formerly the Microsoft Mobile Internet Toolkit (MMIT)—it has further expanded the reach of developers by including the Compact Framework and Smart Device Projects (SDP). These two together allow developers familiar with the desktop Framework to use the same core languages (VB and C#), runtime engine, and IDE to create applications targeted to smart devices, such as the Pocket PC 2002.

But with this increased opportunity come additional challenges for architects and developers. The key is to understand not only the nuts and bolts of the technology, but how and in what scenarios to apply it when building solutions. This is the reason we wrote this book, and we hope the reason you've decided to read it. That is also why we strive in each section to provide the big picture in order to give you enough technical information to understand the challenges and techniques that can be used, while not getting bogged down in every last detail of implementation. For this reason, you'll notice that we rely heavily on short code snippets, listings, and pointers to other resources.

A Note about Audience

As mentioned in the previous section, we've written this book with architects, technical managers, and developers working in the corporate world in mind. Because these three roles often have a large diversity of skill sets, there may be parts of the discussion that are too detailed for some readers. If you find this to be the case, we urge you to rely on the executive summaries at the beginning of each chapter, which summarize key chapter points. We also recommend looking for the key point icons in the margins. These icons indicate special considerations or, as the name implies, key points that we don't want you to miss.

KEY POINT

For other readers the code snippets and listings may leave you wanting more. This is good and done by design; for this reason at the end of each chapter we've included a "Related Reading" section that discusses chapter topics in greater depth. Because some of the entries in the "Related Reading" sections necessarily refer to Web sites, keep in mind that links may change, and you may need to search for the article title, rather than using the address provided. For your convenience, we've placed the links on our site at http://atomic.quilogy.com/cfbook and will do our best to keep them updated.

When considering the skill sets of developers and technical managers reading this book, we specifically had in mind (1) desktop Framework developers and managers who are now embarking on building smart device solutions, and (2) embedded VB developers who are now moving from eMbedded Visual Basic 3.0 (eVB) to the Compact Framework. As you can imagine, the needs of these two groups are inverse to one another. The former require little in the way of managed code and Visual Studio guidance, but more coverage of issues related to developing for mobile devices. The latter are already familiar with building solutions that include mobility but require information on Compact Framework specifics. Because satisfying both audiences is difficult, we've tried to walk a line that balances the two views without leaning too heavily in either direction.

Language Choice

One of the great benefits of working with the .NET Framework is the freedom to work in the language of your choice. This is an inherent benefit of working with code that is compiled first to an intermediate language and then finally to machine code at runtime.

Although the Compact Framework restricts this freedom somewhat through its support for either C# or VB .NET, applications written in either are functionally equivalent and on a par in terms of performance. And as with the desktop Framework, the class libraries that ship with the Compact Framework are the truly interesting part of the product because they encapsulate the functionality that developers will use. All the class libraries are accessible from either language, and so, learning what they have to offer, rather than language-specific syntax, is the key to building both desktop and Compact Framework applications.

For these reasons you'll notice that we included code snippets and listings in one of the two languages (not both, although rest assured that it is possible to translate any C# code to VB .NET and vice versa), although VB .NET is used more frequently to reflect the likelihood that a majority of eVB developers will choose to use VB .NET over C# because of its similarity to the syntax they are familiar with.

The Scope of This Book

This book is organized into three parts: The PDA Development Landscape with the Compact Framework, Essential Architectural Concepts, and Additional Programming Considerations. Each of the parts varies in length, with Part II, which describes the central architectures for handling data, being the largest and consisting of four key topics in five chapters.

The PDA Development Landscape with the Compact Framework

The first part of this book consists of Chapters 1 and 2 and lays the foundation for architecting solutions using the Compact Framework. The goals of these chapters are to put the Compact Framework in the context of mobility (Chapter 1) and to explicate the architecture and core features (Chapter 2) of the Compact Framework and SDP.

Together these chapters are written so that architects and developers will gain an understanding of how the Compact Framework is positioned and why you might want to develop solutions using this technology.

Essential Architectural Concepts

The part on architectural concepts is broken into four key topics that architects and developers typically need to address in their solutions:

1. *Local data handling*. Business solutions that are written using the Compact Framework will typically need to be able to handle relational, file-based, and XML data in order to display it to the user. Chapter 3 addresses this need by discussing how applications can access local data using the support in the Compact Framework for processing XML, reading and writing to files, and using ADO.NET. The techniques in this chapter apply to handling disconnected data, as well as data downloaded from a remote source.
2. *Remote data access (RDA)*. Business solutions also have the need for accessing data over the Internet or on corporate networks. Chapter 4 addresses this requirement by discussing various architectures for accessing remote data including XML Web Services, sending HTTP requests, low-level networking using sockets and infrared standards developed by the Infrared Data Association (IrDA), and accessing Microsoft SQL Server 2000 remotely. The techniques in this chapter apply to connected scenarios.
3. *Robust data caching*. Chapter 5 discusses how applications can cache data locally on a device using SQL Server 2000 Windows CE Edition 2.0 (SQLCE 2.0). Most of the discussion will focus on the uses for and features of SQLCE 2.0 because it has many advantages over simply caching data through the use of files.
4. *Synchronization*. Finally, Chapters 6 and 7 discuss ways in which data can be synchronized, including a discussion of simple synchronization with a host PC using ActiveSync (Chapter 6) and following with the RDA and merge replication features of SQLCE 2.0.

Of course, all of these topics reflect the centrality of data access. This is no coincidence, because architecting and developing business solutions involve, in large measure, moving data around, editing it, saving it, and synchronizing it. As a result, in our discussion of the third and fourth concepts mentioned here, we focus considerably on SQLCE 2.0 and how it can be used with the Compact Framework.

This book does not address architecting solutions using Microsoft ASP.NET Mobile Controls (formerly MMIT)—a set of ASP.NET server controls that enables the development of Web applications for a variety of mobile devices—because there are other good resources already available and because the kinds of applications developed with Mobile Controls are fundamentally different from those developed with the .NET Compact Framework.

Additional Programming Considerations

The third part of the book consists of three topics that most mobile solutions will need to address: localization, security, deployment, and developer challenges.

Chapter 8 addresses how Compact Framework applications can be architected to adapt to devices running in different geographies. Chapter 9 is dedicated to the various techniques for securing both the device on which the Compact Framework code runs as well as its data. Chapter 10 discusses techniques for deploying Compact Framework solutions. Finally, the book concludes with a single chapter that discusses some of the challenges developers will face when building solutions with the Compact Framework. Many of these challenges relate to implementing features that are not directly supported by the Compact Framework or that are somewhat difficult to grasp at first.

Final Words

In addition to our consulting and teaching activities, we write technical articles (including some on the Compact Framework) for our Web site at http://atomic.quilogy.com. We encourage you to check out the site and hope you'll use it as a technical resource.

In the end our hope is that you will feel the time and money invested in this book have been worth it because you have taken away important concepts and techniques that you can apply as you build great solutions using the Compact Framework. If you have questions or comments please feel free to e-mail us at atomic@quilogy.com.

Dan Fox
Shawnee, Kansas

Jon Box
Memphis, Tennessee

October 2003

Acknowledgments

Book projects are, almost without exception, team efforts, and this one was no exception. First, we'd like to thank Sondra Scott, senior acquisitions editor at Addison-Wesley, who helped shape the initial idea for the book and has been there through every step in the process. One of us has known Sondra for some time, having worked with her on other projects, and can say truthfully that she's always kind, professional, and helpful. What more can authors ask?

We'd also like to thank the other folks at Addison-Wesley who helped by providing feedback on the manuscript and coordination with technical reviews and production, including Jennifer Kelland (who did her best to make most of our writing at least intelligible), Emily Frey, Jennifer Blackwell, and Marcy Barnes. We also appreciate the several cover designs created by Karin Hansen and her patience as we reviewed them and asked for modifications. To any others at Addison-Wesley that we've missed, we'd like just to offer a general thank you.

Technical books need good technical reviewers, and we have two. Kent Sharkey from Microsoft assisted us with the book even before the writing stage, offering valuable insights and validation that the relevant material has been covered. Michael Yuan's feedback has also been valuable; more than a few paragraphs in this book began as ideas that Michael gave us. From a technical perspective, several talented individuals contributed by reviewing the original concept for the book, including David Chun, Joel Mueller, Chris Muench, Larry Wall, Doug Ellis, Scott Bechtold, and Don Michael.

We'd also like to thank several of our coworkers at Quilogy (the IT services organization for which we work) who have encouraged us and otherwise assisted with the book in a variety of ways. These start at the top with the president and CEO, Randy Schilling, for his indulgence while we worked on the book, but also include Manish Chandak, who has had to juggle schedules and assignments during the process, and Deborah Moulton, who provided great feedback on cover design and has helped promote the book. Quilogy is a team, and so we'd also like to thank the rest of the organization, whose professionalism and skill provide the foundation from which we were able to complete this project.

Finally and most important, we would like to thank our families: Beth, Laura, and Anna; Tonya, Stephen, Taylor, and Trace. Without your support and encouragement, not only would the book never have been completed, it wouldn't have been worth finishing. Thank you.

About the Authors

Dan Fox is a Technical Director for Quilogy (http://www.quilogy.com) in St. Charles, Missouri. Quilogy is a leading Microsoft Gold Certified Partner and Solutions Provider. Within Quilogy, Dan is a member of the Atomic team that focuses on emerging technologies by providing consulting and custom technical education. Within the Atomic group Dan has coauthored and delivered the Atomic .NET 4-day course on multilayer Web development and the Atomic .NET Patterns and Architecture 3-day course for architecting applications with the Windows .NET Framework. The course syllabi and schedules can be found on http://atomic.quilogy.com.

Dan is also the coauthor of *MCSD Exam Cram Visual Basic 6 Distributed* from Coriolis and author of *Pure Visual Basic*, *Building Distributed Applications with Visual Basic .NET*, and *Teach Yourself ADO.NET in 21 Days* from Sams. Since 1998 he has authored more than two dozen articles for *Visual Basic Programmer's Journal*, *Visual Studio Magazine*, *.NET Magazine*, *SQL Server Magazine*, *Advisor Journals*, *MSDN Magazine*, Builder.com, and InformIT.com. He is also a member of the Microsoft author community.

Dan has spoken at numerous events, including Visual Studio .NET launches (2002), Developer Days (1997–2001), .NET Architect Summit 2001, and TechEd (2001, 2001 Europe, 2002). His focus has been on development on the Microsoft platform specializing in Web development and data access technologies. He has served as a trainer, consultant, and managing consultant with Quilogy since 1995 in its Kansas City office.

Dan lives in Shawnee, Kansas, with his wife, Beth and two daughters, Laura and Anna, where he waits and optimistically hopes that for the Cubs, next year is finally here.

As a Solution Architect at Quilogy, **Jon Box** serves as a .NET evangelist, focused on helping developers utilize .NET. Jon's current work emphasizes mobility, using the Compact Framework and ASP.NET Mobile Controls.

Jon has been programming professionally since 1985. In that time, he has worked in a variety of environments and languages that include Cobol,

Assembler, Clipper, C, C++ (ATL, MFC, Win32), VB, VB.NET, and C#. Jon is completing his fourth year at Quilogy and has had many responsibilities in that time including instructor, developer, project manager, managing consultant, and general manager.

One of Jon's proudest affiliations is that of being a Microsoft Regional Director, which is a program of Microsoft unpaid evangelists who are nominated by Microsoft field employees. This program has about 30 members in the United States and another 100 around the world. Having been nominated for the third year, Jon considers being a part of this group an honor and privilege. See http://www.microsoft.com/rd for more information about the program.

Currently, Jon is in his second year as a member of the Atomic group at Quilogy. Not only did he present the original idea for the group's existence, he cowrote with Dan Fox the Atomic .NET class which is a solution-oriented training for ASP.NET development and is presented all over the United States. At Quilogy Atomic continues to be a group of top presenters and are considered part of the company's technical thought leadership.

Jon began writing while at Quilogy. He started by being the technical reviewer of *An Introduction to Object-Oriented Programming with Visual Basic .NET* from Apress. Since then, Jon has developed several magazine articles, articles for the Atomic Web site, and white papers for Microsoft's Smart Devices Developer Community (http://www.microsoftdev.com). Jon coauthors with Dan Fox the Mobility column for *.NET Developer's Journal* and is a member of the magazine's editorial board (http://www.sys-con.com/dotnet).

Jon presented sessions on .NET Performance and Data Access for Sometimes Disconnected Devices at Tech·Ed 2003. He also has presented a variety of MSDN WebCasts on topics such as asynchronous programming, ASP.NET Mobile Controls, and the Compact Framework.

Jon founded the Memphis .NET User Group (http://www.memphisdot.net) and is also a member of the INETA Speakers Bureau (http://www.ineta.org). Ask your local .NET User Group leader to contact INETA about scheduling a presentation with some of the best .NET minds in the universe.

Jon resides in Memphis, Tennessee, with his wife, Tonya, and sons, Stephen, Taylor, and Trace. When not working or asking Dan questions, Jon can be found near a baseball field watching his sons play, coaching the Memphis Eagles, or checking on how the St. Louis Cardinals are doing in the NL Central.

PART I

The PDA Development Landscape with the Compact Framework

CHAPTER 1

The Rise of Mobile Devices

Executive Summary

Today we are witnessing an explosion in the use of mobility in both our personal and work lives. In the years ahead most workers will rely on mobility every day, whether on the road or in a typical office environment.

As a result, people will need to access information anywhere, anytime, and on any device. This includes corporate infrastructure (e-mail and calendaring), custom line-of-business applications, and information that exists in packaged horizontal applications. Workers also need to access the information when disconnected from the corporate network, as well as when using devices that function in connected and occasionally connected modes. In addition, there is a vast array of devices from PDAs to wireless Web phones that architects and developers must consider when architecting their solutions.

This book concentrates on a subset of that entire picture by focusing primarily on architecting line-of-business applications for PDAs using the Microsoft Windows .NET Compact Framework.

Microsoft has a history with mobility stretching back over ten years, most of it squarely centered on the Windows CE operating system. The .NET Compact Framework and Smart Device Projects (SDP) for Visual Studio .NET 2003 target Pocket PC 2000, 2002, and embedded Windows CE .NET 4.1 platforms. This toolset produces managed code using a subset of either Visual Basic (VB) .NET or Visual C# .NET that is executed by a common language runtime analogous to the desktop Microsoft Windows .NET Framework.

The Need for Mobility

The word "mobility," meaning the "capacity for motion," has been in English usage since at least 1425 (derived from the Middle French and ultimately from the Latin *mobilitatem*). However, the word has taken on new meaning in recent years as the interest in equipping both consumers and business users with "information anywhere, anytime, and on any device" has gained momentum. In fact, this mantra is in part lifted from Microsoft's vision statement that changed from Bill Gates and Paul Allen's original 1975 vision of "a computer on every desk and in every home." The new statement, "empower people through great software anytime, any place and on any device," was created in mid-1999 as Microsoft caught hold of this trend in mobility, and that ultimately resulted in you reading this book today. The Compact Framework is one part of Microsoft's vision for mobility. Microsoft likes to call this new and exciting period the PC-plus era.

This trend can be traced from the initial adoption and subsequent spread of cellular technology in the 1980s to the personal digital assistant (PDA), wireless Web phones, and smartphones of today. This technology has in turn helped transform not only business culture, but also our entire society into one that values and depends on mobility.

NOTE: In 2001, an estimated 300 million wireless phones were used worldwide. This number is expected to grow to one billion by 2005.

At a societal level this capacity for motion has allowed people to integrate their work and home lives to a greater degree than ever before by managing the ever increasing flow of information through the use of these electronic tools. A recent study indicated that today, 19% of workers describe themselves as mobile workers. Clearly, the culture of mobility is now ingrained in today's workers, a fact that is clearly of interest to software developers and architects as they contemplate serving business users in the coming years.

And in fact, if industry forecasts are any indication, the adoption of mobile technologies and mobile life and work styles will only increase in the years ahead. For example, the research firm IDC in July of 2002 estimated that the number of mobile workers in the United States will increase from 92 to 105 million from 2001 to 2006, that segment growing twice as fast as the general workforce, while they estimate that 60% of Fortune 1000 companies will deploy a mobile application server by the end of 2003. However, this trend isn't confined to the largest organizations. In fact, a recent survey indicates that nearly half of small and medium-sized businesses are making wireless devices a spending priority.[1] The end result is that by 2006, two-

---

[1] VAR Business.com, "IT Strategies in SMB," August 2002.

thirds of all workers will be mobile in some capacity, which includes everything from traditional "road warriors" to workers roaming with a wireless device on a corporate campus.

As you might imagine, the combination of the acceptance of mobile work styles and more advanced technology has created a positive reinforcement pattern. As the demand for mobile technologies increases, the market responds with technological advances, and it is those very advances that drive demand for yet more sophisticated and integrated technologies. Several recent technologies that have participated in this feedback loop to fuel more productive applications include XML Web Services; wireless LAN (WLAN) technology, including the Institute of Electrical and Electronics Engineers (IEEE) 802.11 specifications; Wireless Application Protocol (WAP) for wireless Web phones; Virtual Private Networks (VPNs); wireless personal area network (WPAN) technologies, such as BlueTooth; the anticipated rollout of 2.5G and 3G networks by major carriers; and more sophisticated devices like the Pocket PC 2002 and 2003 with built-in support for streaming media, the Tablet PC, and smartphones; and, of course, robust development and runtime environments like the Compact Framework.

Further, in many cases, the technologies themselves are entangled within the loop. For example, as the number of business workers with wireless Internet capability (estimated by the IDC in November 2001 to increase from 2.6 million in 2000 to 49 million by 2005) increases, there will be a corresponding increase in the investment in public WLAN technology,[2] where high-speed wireless networks are available in public spaces such as airports and malls for use by business workers and the public at large.

This positive feedback loop will likely continue into the future as new technologies, such as Mobile IP and the Multimedia Messaging Service (MMS), are developed that enable multimodal communications to take advantage of 3G networks allowing users to exchange photos, sounds, video clips, and text messages. For a good overview of the technological landscape for wireless computing now and into the future see the book by Alesso and Smith in the "Related Reading" section at the end of the chapter.

Of course, there is more behind the need to create mobile applications than simply a chase-your-tail mentality where businesses adopt technology for technology's sake. By providing anytime, anywhere access to information, a business can immediately realize two primary benefits that have cascading implications.

[2] Estimated by Cahners In-Stat/MDR in January 2002 to grow almost 60-fold by 2005.

1. *Making faster and more informed decisions:* If business users are able to access corporate information (e-mail, corporate documents, line-of-business data) on mobile devices and not only when they are officially in the office, they make decisions in a timely fashion. This often results in heading off problems, leading to increased revenue and greater customer satisfaction. The motivation behind faster decision making and more empowered employees has been proven many times over in the implementation of corporate Intranet portals where a cross section of job-specific information is accessible to business users, thereby enabling them to collate all the relevant data.
2. *Improved data accuracy:* Many of the first mobile applications businesses implement automate processes involving data collection. Using a mobile device, business users can capture electronically information that in the past involved handwritten forms, followed by faxing and data entry. This leads not only to information that is accurate, but also to reduced cycle times and a significant labor and cost reduction.

There is now a synergy of business rationale, culture, and technology that will enable business to create robust and exciting mobile applications. In the next section we'll unpack the concept of information anywhere, anytime, and on any device to look at several of the common mobile scenarios, along with the types of devices that are used.

Information Anywhere, Anytime, and on Any Device

The research and advisory firm Gartner has said that 25% of enterprises have developed a cohesive strategy for extending their applications to mobile workers. As these businesses (and hopefully yours) move to incorporate mobility, they often do so in the context of a few well-defined scenarios using a set of mobile devices with particular characteristics. In this section we'll briefly explore these scenarios and devices by unpacking the concept of accessing information anywhere, anytime, and on any device.

Using Information

When you think of providing access to information anywhere, anytime, and on any device in the business world, you can divide that information into three primary categories: corporate computing infrastructure, line-of-business applications, and horizontal application areas. These areas cut across all the primary verticals, including financial services, retail, government, health care, and manufacturing.

Corporate Computing

For businesses that have embraced mobility, one of the crucial components is extending the corporate computing infrastructure to mobile devices. This is accomplished by giving access to corporate resources such as e-mail and scheduling, file servers for documents, and Intranets to mobile workers through the use of WAP-enabled phones and VPNs on laptops and PDAs. Various IT vendors have helped to support this extension by providing server products such as Microsoft's Mobile Information Server (MIS) in 2001 and enabling synchronization of Microsoft Outlook data to PDAs. As creating the infrastructure becomes more mainstream, these vendors will roll mobility services right into core products, as Microsoft is doing by rolling the functionality of MIS into a future release of Exchange Server, their core messaging product.

Line-of-Business Applications

Every medium- to large-sized business has developed some custom line-of-business applications that support their core activities. These can include everything from on-site inspections to order entry to shop floor scheduling to a variety of data-entry systems. By extending these applications to mobile devices, businesses can benefit from more timely access to information, more accurate data, and streamlined processes.

Horizontal Applications

There are also several horizontal application areas that businesses gravitate toward because they naturally have a large impact on customer interaction and revenue. These include the following:

- *Customer relationship management (CRM):* Sales representatives, account executives, field service technicians, and customer support users all require timely and accurate access to customer information. A CRM system that is available via a mobile device can assist these users by providing customer contact information, history, buying trends, and product information that lead to increased revenue. Vendors such as Siebel Systems and Clarify provide packaged CRM systems.

- *Enterprise resource planning (ERP):* Businesses that have implemented ERP packages from vendors such as SAP AG and J. D. Edwards and Company will often look to provide this information to mobile users. Having access to this information when away from the office can improve productivity and enable quicker responses.

As you look at the types of information that businesses aim to provide, the sweet spot for the Compact Framework, and therefore for this book, is line-of-business applications.

In many instances the corporate computing infrastructure will be serviced through IT vendors' server-based products, such as Microsoft MIS and Web-based solutions, while commercial CRM and ERP packages may include a wireless component. However, the custom business processes, data, and logic underlying most line-of-business applications are incorporated in software that must be adapted or rewritten to accommodate mobile devices. The Compact Framework is ideally suited to this job, as we'll discuss in Chapter 2.

Anywhere and Anytime

The second component of mobility involves accessing information anywhere and anytime. Essentially, this formulation addresses the issue of connectivity with a network and if data is persisted on the device.

At the most basic level, a mobile device can be used in one of three modes: disconnected, connected, and occasionally connected, all of which are addressed by the Compact Framework and associated tools.

- *Disconnected:* In this mode the device is not connected to a network, and applications must therefore rely on the resources and data already on the device. Examples of disconnected applications might include utilities and games, a note-taking application, and an application to keep score at a baseball game. Of course, the capabilities of the device govern whether applications can be run successfully in disconnected mode. For example, a PDA has local storage and an operating system that supports executing custom code, whereas wireless Web phones typically do not have that capacity. In Chapters 3 and 5 we'll discuss the options for working with locally persisted data.
- *Connected:* In this mode the device is connected to a network, and the application requires the connection in order to function. In the case of a PDA, this can be accomplished in a variety of ways, for example, through a universal serial bus (USB) cable attached to a PC,

a WLAN card using an 802.11 protocol, a General Packet Radio Service (GPRS) card, or even through integration with a phone, as we'll discuss shortly. For wireless Web phones, this means using WAP through a WAP gateway to access Web content in the form of Wireless Markup Language (WML). In either case, in this mode, applications, such as instant messaging, conferencing, real-time inventory, or stock trading, running on the device can retrieve and update information in real time using either proprietary protocols or industry standards, such as XML Web Services. The accelerating adoption of high-speed wireless technologies coupled with XML Web Services makes this mode the fastest growing of the three; Chapter 4 will explore the topic in greater detail.

- *Occasionally connected:* This mode probably includes the majority of mobile applications being developed by businesses today. In this mode, the device is sometimes connected to the network but must work whether connected or not. The typical scenario is a home health care application where, before leaving the office, the user synchronizes a PDA with data, including the visit schedule and patient information for that day. While at the patient location, no reliable network connection exists, and so the application must store data locally until it can be uploaded to a central repository later in the day. These scenarios are often more complicated and will be discussed in Chapters 6 and 7.

On Any Device

The final aspect of mobility is accessing information on any device. Although there have been a wide range of mobile devices developed over the past few years, they generally fall into the categories shown in Figure 1-1.

Laptops and Tablet PCs

Laptops and notebook computers were the original mobile devices, running Intel processors and running Windows software. These devices can access wireless networks, for example, by using 802.11 networking and wireless PC cards, and can access Global System for Mobile Communication (GSM) and GPRS networks using wireless modems. While these can be thought of as the original mobile devices, the enterprise purchases of notebook computers is projected to increase by 40% in North America, 100% in Europe, and 200% in Asia over the next three years.[3]

---

[3] Meta Group, August 2002.

Figure 1–1 *Mobile devices.* This diagram shows the different types of mobile devices and their corresponding categories.

The latest twist is the Tablet PC, introduced in late 2002, a thin, keyboardless device that runs a superset of the Microsoft Windows XP operating system and includes voice and handwriting recognition, in addition to being 802.11 ready, making this device ideal for roaming business users in a campus setting.

Wireless Web Phones

With the explosion of the Internet in the mid-1990s, cellular phone manufacturers and service providers such as Sprint PCS, AT&T, Nextel, and Verizon began to look for ways to incorporate Web technologies into their products. Some of the protocols that were developed included Handheld Device Markup Language (HDML) from Unwired Planet, Tagged Text Markup Language (TTML) from Nokia, and Intelligent Terminal Transfer Protocol (ITTP) from Ericsson. However, the market began to coalesce around the WAP protocol with its WML standard, developed in mid-1997.

NOTE: Although WAP and WML are more popular in the United States, protocols and standards other than WAP persist and enjoy wide support, including I-mode (popular in Japan), HDML, compact HTML (cHTML), and its superset Mobile Markup Language (MML).

So-called wireless Web phones are now available from all the major manufacturers and use wireless Web Services through the service providers mentioned previously. These devices have helped to fuel the rapid increase in the use of mobile data services, expected to grow by 50% per year through 2005.[4]

Wireless Web phones are typically used for accessing corporate infrastructure, such as e-mail, calendaring, expense reporting, and other Intranet functionalities. Their lack of local storage means they are used in a connected mode.

PDAs

Personal Digital Assistants is a general term used for mobile devices that typically have some local storage capabilities and include calendaring, contacts, and task management. These devices differ from PCs because they typically use lower-power central processing unit (CPU) architectures (such as ARM, SH, XScale, and MIPS) and operating systems specially designed for mobile devices. Customized applications can be written for these devices and downloaded and executed. In addition, they ship with the capability to synchronize data with a PC and can typically be extended for wireless access through expansion cards. As a result, they can be used in connected, disconnected, or occasionally connected modes.

Although different people use different terminology, PDAs generally fall into two distinct types:

1. *Palm:* These devices are small in size (roughly 3" × 5" in portrait mode) and receive their input via a touch screen and stylus, using either handwriting recognition or specific strokes with the stylus. Examples include the 3COM's PalmPilot and the Pocket PC 2000 and 2002 devices, such as Compaq's iPAQ, and Casio's E-200. Palm devices typically have longer battery life than handheld devices.
2. *Handheld:* These devices are differentiated by a clamshell design (many approximately 3.5" × 7" in landscape mode) that includes a keyboard, in addition to a touch screen and larger displays, which

[4] Giga Information Group, June 2002.

make the largest of them only slightly smaller than a traditional notebook or laptop PC. Some of these devices include the NEC MobilePro 880 and Hewlett-Packard (HP) Jornada 820.

These devices have obviously been critical to the adoption of mobility, and their usage will continue to grow. Forrester Research estimated in 2001, for example, that PDA usage would increase by 267% in Global 3500 companies from mid-2001 to mid-2003.

Smartphones and PDA Phones

More recently, the concepts of the PDA and mobile phones have converged in so-called smartphones, and PDA phones. Basically, these two types of devices incorporate the functions of both a PDA and a mobile phone, although the emphasis of each differs. In the case of the smartphone, the form factor of the device is essentially a wireless Web phone with a slightly larger display and the possible inclusion of a stylus. On a PDA phone, the form factor is that of a PDA with the inclusion of an antenna and the ability to make calls by tapping contacts or phone dialer software on the device.

NOTE: In late 2002, Microsoft released its SmartPhone 2002 software development kit (SDK) with plans for OEMs to release first-generation smartphone devices by the end of 2002. The first device announced to support the platform was the Orange Sounds Pictures Video (SPV) from London-based Orange SA, providing service in the United Kingdom and Europe beginning in November 2002. AT&T Wireless also announced plans to release smartphone-based devices for use in the United States.

A recent example of the latter is the T-Mobile Pocket PC Phone Edition offered by T-Mobile and VoiceStream Wireless, which uses the T-Mobile GSM/GPRS high-speed wireless voice and data network. These Pocket PC Phone Edition devices ship in two configurations, using different frequencies for Europe and the United States.

Because they offer a network connection and local storage, both smartphones and PDA phones can be used in connected, disconnected, or occasionally connected modes.

Special-Purpose Devices

Finally, there are also mobile devices that are manufactured for more specialized purposes. These devices can adopt either the standard Palm or handheld form factor and simply add features, such as a "ruggedized" case for surviving dropping and outdoor usage (e.g., the UltraPad 2700 from Ameranth), or depart from these for specialized purposes such as data collection and bar code scanning in devices like the Intermec 5020 Handheld.

In fact, these special-purpose devices can also be embedded within a larger device, an early example includes the Clarion AutoPC, which fits into the dashboard of a car and is controlled through voice recognition. In addition, many car companies are currently developing specialized embedded systems for their vehicles that are collectively referred to as *telematics*. For example, in March of 2002 it was announced that BMW used embedded Microsoft Windows CE technology in its BMW Series 7 line of cars in the Control Display component of what BMW calls iDrive.

There are also devices that may be embedded into, and are used to, control a variety of other mechanical devices and appliances from vending machines to toasters and that are generally referred to simply as embedded devices.[5]

KEY POINT

At this time there is no single development platform or environment that targets this large range of devices. Fortunately, the Compact Framework allows you to write software for a fairly large and growing segment of these devices, including Palm devices like the Pocket PC 2000 and 2002 and PDA phones like the Pocket PC Phone Edition (with some caveats) and some embedded devices that run the new Windows CE .NET 4.1 operating system. It does not, however, support any of the handheld or smartphone devices, although look for support for smartphone devices in a future release.

In the following section we'll outline Microsoft's involvement in mobility and see how it ties into the Compact Framework.

Microsoft and Mobility

Microsoft has been investing in mobility for over a decade. This investment began in 1992 with two projects code-named Pulsar and WinPad. The WinPad project was initially designed to produce a handheld PC (then termed a "PC Companion") to compete with Apple's Newton MessagePad (which itself eventually failed to catch on) based on Windows 3.1, while the Pulsar project was aimed more at pager-like devices using an object-oriented operating system. However, due to internal struggles as to the feature sets to

[5] Additional categories of devices that have not gained wide acceptance include Web companions, which are essentially Internet appliances for home use, and Web-enabled phones, which are wired phones with an attached Internet terminal.

support and hardware issues, these projects ultimately failed to produce viable products. However, shortly thereafter in 1994, these projects were reorganized under the code name Pegasus, which eventually led to the release in 1996 of Microsoft's first operating system designed for mobile devices, Windows CE.

Although you might think that the CE in Windows CE stands for "Compact Edition," Microsoft insists that CE is not an acronym, but simply the name of the operating system. Microsoft now refers to devices running Windows CE as simply "Windows-powered devices."

Since the release of Windows CE, Microsoft has continued to invest in operating systems, platforms, and development tools for mobility. In this section we'll explore those three concepts in order to put the Compact Framework in context. Note that these concepts are naturally hierarchical with operating systems at the highest level and development tools at the lowest.

Operating Systems

The operating systems that Microsoft has developed in this space can be differentiated between the various versions of Windows CE and embedded operating systems used in specialized devices, ranging from Internet appliances to manufacturing equipment.

Windows CE

Windows CE itself is based on its desktop cousins and, as such, supports an API familiar to Windows developers. As a result, while it is relatively straightforward for developers accustomed to developing desktop applications to move to Windows CE development, there are still considerations that developers must take into account when developing for smaller devices, including the user interface (UI), power consumption, limited storage, and communication options.

In general, the design of Windows CE adhered to the following design principles:

- Small memory footprint
- Modular approach
- Processor portability
- Win32 compatibility
- Connectivity
- Real-time processing

The modular approach allows Windows CE to be broken into about 300 modules (in the latest Windows CE .NET release) from which an OEM can choose to build an operating system. The smallest of these systems can fit in about 200K of RAM, while a typical configuration that includes the graphical interface and applications is about 4MB.

A diagram that illustrates the architecture of Windows CE is shown in Figure 1-2.

As previously mentioned, Windows CE 1.0 was launched in 1996 and was targeted for handheld PDAs (referred to as Handheld PCs or H/PCs) that had small touch-type keyboards, grayscale displays at 480 × 240 resolution, and alkaline batteries. These devices were manufactured by a variety of vendors, including Casio (Cassiopeia A10), Compaq (Companion C120), HP (320LX), and Phillips (Velo I), among others. They supported synchronization of

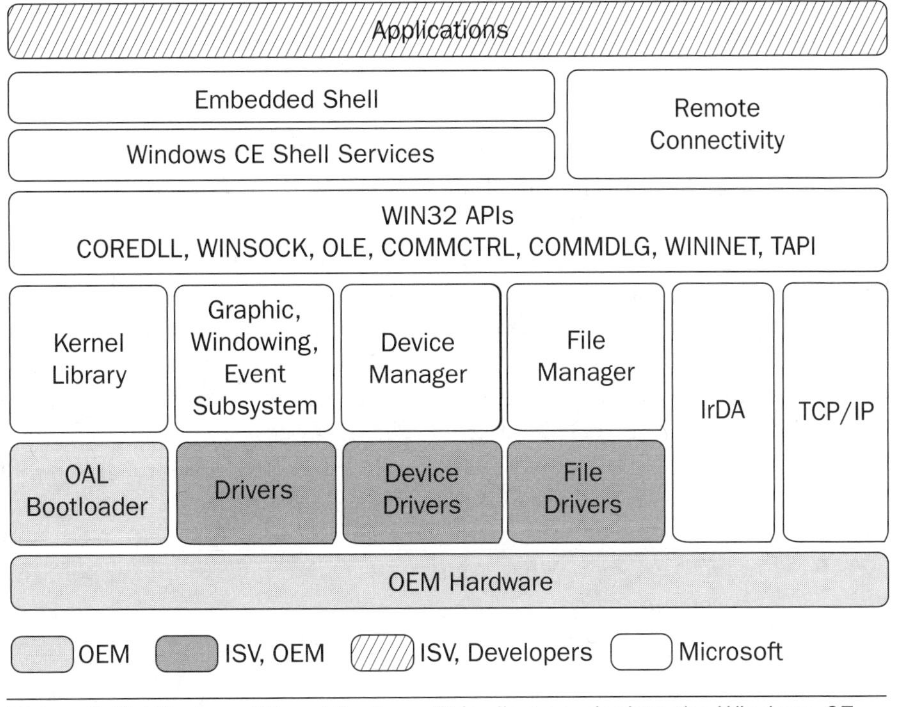

Figure 1–2 *Windows CE architecture*. This diagram depicts the Windows CE architecture. Note that different parts of the architecture are implemented by Microsoft, the OEM, ISVs, and application developers.

Table 1–1 Windows CE Versions

Release	Date	Features
2.0 (Alder)	Fall 1997	Support for demand-paging of code, ActiveX, WinInet API, cascading menus, Ethernet networking, true-type fonts, and x86 processors
2.1 (Birch)	May 1998	Support for USB, NTLM authentication, cryptography APIs, color displays
3.0 (Cedar)	June 2000	Support for Component Object Model (COM) and Microsoft Message Queue (MSMQ), smartcard support, embedded Web server with ASP support
CE .NET 4.0 (Talisker)	January 2002	Automatic configuration of 802.11 networks, BlueTooth, customizable UI elements, Direct3D, DirectMusic APIs, event-based notifications, Kerberos, and many others
CE .NET 4.1 (Jameson)	July 2002	Support for IPv6, file viewers, and additional performance
CE .NET 4.2 (McKendric)	April 2003	Enhanced real-time processing, better performance, enhanced multimedia capabilities, support for IPSec

e-mail and calendaring with Schedule+ and Outlook using a serial connection, as well as expandability via PCMCIA or PC card slots.[6]

Since 1996 Microsoft has continued to enhance the Windows CE operating system. Table 1-1 briefly describes the timing and features of the major releases.

Windows CE was designed to support lower-power CPU architectures that are more suited for mobile devices. These include Advanced RISC Machines (ARM), Microprocessor without Interlocked Pipeline Storage (MIPS), PowerPC (another RISC processor), XScale (from Intel), and SuperH (SH) from Hitachi, another RISC architecture). With each release of Windows CE, the number of supported processors from various vendors such as ARM, NEC, Toshiba, Hitachi, and Intel has increased, with the notable exception that Windows CE .NET does not support any PowerPC processors.

[6] One of the authors of this book fondly remembers the day he bought his Cassiopeia A10 and joined the world of mobile computing. The device served him well for about two years, until he inadvertently left it on top of his car while backing out of the driveway. Needless to say, the device was not “ruggedized” and, even if it had been, would likely not have survived the 30 MPH impact with the curb. Such are the dangers of mobile computing.

All of the Windows CE versions also ship with productivity and supporting software, including Pocket Office (Pocket Word, Pocket Excel, and later Pocket PowerPoint and Pocket Access), Pocket Outlook, Schedule+ (in CE 1.0), Pocket Explorer, and synchronization software that in its first incarnation was called Handheld PC Explorer and later ActiveSync 3.1 and 3.5. The Pocket PC can also run a version of Windows Media Player and Microsoft Reader, in addition to other software you can download from www.microsoft.com/mobile/pocketpc/downloads/default.asp.

Embedded

Although Microsoft eventually released Windows CE, work continued on a scaled-down version of a Windows operating system that could be embedded in specialized devices, such as set-top boxes, kiosks, retail point-of-sale (POS) systems, Internet appliances, and manufacturing equipment. These efforts eventually led to two embedded operating systems, Windows NT Embedded 4.0 in 2000 and Windows XP Embedded in 2002. Essentially, these are modularized versions of the desktop operating systems.

Typically, OEMs would choose an embedded operating system when careful power management and a footprint of under 5MB are not required, when using an x86 architecture, when support for existing Win32 applications is required, and when extensive networking and communications support is required. For example, OEMs can use the Windows Embedded Studio tool to build a platform based on Windows XP Embedded by choosing from the available modules or using a set of predefined templates, including three for set-top boxes, a home gateway, kiosk, gaming console, network-attached storage (NAS), POS, and information appliances.

NOTE: In addition, Microsoft offers a Server Appliance Kit (SAK) for Windows 2000 designed to help OEMs design server appliances such as Web servers, NAS devices, and other customer appliances.

Microsoft usually uses the term "embedded" to refer to both Windows CE .NET and Windows XP Embedded and, therefore, provides information on its Web site for both at www.microsoft.com/windows/embedded.

At this time the Compact Framework does not support developing software on either of the embedded operating systems from Microsoft.

Platforms

It's important to keep in mind the distinction between the operating systems discussed in the previous section and what Microsoft refers to as "platforms." Essentially, a platform is defined by the combination of a specific set of hardware, a set of programs, modules, UI components, and an operating system. A platform defines all the key characteristics of a class of devices in a single package, so that software developers and device manufacturers can more easily target their solutions.

For example, in the case of the Pocket PC platform, Microsoft defined the hardware specifications that vendors such as Compaq and HP implemented. The platform included the Windows CE 3.0 operating system in addition to extensions such as the Pocket PC shell and Notification API. For software developers, all of these APIs are packaged into a platform SDK available on Microsoft's Web site. Table 1-2 shows some of the platforms based on Windows CE.

Table 1–2 Platforms Developers Target Using the Platform SDKs from Microsoft

Platform	Operating System	Notes
Handheld PC	CE 1.0, 2.0	Code-named "Mercury." The original H/PC devices such as the Cassiopeia A10 with 2MB to 8MB of RAM.
Handheld PC Professional	CE 2.11, 3.0	Originally introduced in 1998 and reissued as Handheld PC 2000 in 2000. Used primarily in larger keyboard-based devices (10" × 8") with 16MB or 32MB of RAM and a wide variety of specifications. Includes Pocket Access. Examples include HP Jornada 720 and Intermec 6651. These come in half-screen (code-named "Callisto") and full-screen (code-named "Jupiter") devices.
Auto PC	CE 2.0, 3.0	Code-named "Apollo" and first available in 1998. Since then, the Auto PC has been replaced by the Windows CE for Automotive* platform for telematics and the more expansive "Car .NET" initiative.
Palm-Size PC (P/PC)	CE 2.1, 2.11	Code-named "Gryphon," available in 1998, and included devices such as the Compaq Aero. Designed for monochrome devices at 320 × 240 with 2MB to 16MB of RAM. These devices were quickly made obsolete by the Pocket PC platform.

*See www.microsoft.com/automotive/ for more information on Windows CE Automotive version 3.5 and Car .NET.

Platform	Operating System	Notes
Pocket PC 2000	CE 3.0	Code-named "Rapier" and abbreviated PPC. This platform was based on an early version of CE 3.0. The Compaq iPAQ is the most popular device on this platform that can be thought of as a superset of the Palm-Size PC platform with similar specifications but with faster processors and more memory and battery life. These devices use a variety of processors with 32MB of RAM. Devices that support FlashROM, such as the Compaq iPAQ, can be upgraded to Pocket PC 2002 because they support FlashROM.
Pocket PC 2002	CE 3.0	Code-named "Merlin" and built on the fully released CE 3.0. Must use the Intel StrongARM (or compatible processor such as the Intel X-Scale) with 32MB or 64MB of RAM. Must also include 32MB FlashROM for system software and must support upgrades and use a hybrid transflective thin film transistor (TFT) screen. The Pocket PC 2002 Phone Edition (code-named "Space Needle") is a special version of this platform. The Intermec 700 and Compaq iPaq support this platform.
Pocket PC 2003	CE .NET	Code-named "Peregrine" and will include devices with a square screen and a keyboard and some with a portrait screen and no keyboard. Not yet released at the time of this writing.
SmartPhone 2002	CE 3.0	Released in late 2002 and code-named "Stinger." Targeted at phones also supporting traditional PDA features, including color screens, synchronization, local storage, and applications. The first announced device was the Orange SPV in October 2002.
Tablet PC	Superset of Windows XP called Windows XP Professional Tablet PC Edition	Device available Fall 2002. Portrait-based, keyboardless device with high-resolution displays, voice, handwriting recognition, and 802.11 networking
Smart Display Monitor	CE .NET 4.1	Code-named "Mira" and includes display devices targeted for home users that enable them to view their Windows XP Professional PC from elsewhere in the home using 802.11 technology.

Screen Technology

Color LCD displays come in two types, active and passive. Active displays such as thin film transistor (TFT) offer a sharper image by refreshing the screen more frequently than passive displays, although newer, passive technologies, such as color super-twisted nematic (CSTN) developed by Sharp, are similar to active displays but are half the cost.

LCDs also support two types of lighting, transmissive and reflective. Reflective screens use much less power because they use external light, such as light around the device or a front-lighting system, but are darker when viewed indoors. Transmissive lighting provides backlighting from a light source behind the screen, but this causes the display to be unreadable outdoors.

It should also be noted that both Windows CE 3.0 and Windows CE .NET also include a platform builder that allows vendors to adapt Windows CE to their hardware. Whether adapting their hardware to an existing platform, such as Pocket PC 2002 (by, for example, mapping buttons on the device to functions in the operating system), or building an entirely new platform for an embedded system, OEMs can use the tools to build a modularized Windows CE operating system by choosing from the 300 modules in Windows CE .NET. Using the platform builder (and the associated wizards), OEMs can define which processors the system supports and choose which of the components to include, in addition to providing low-level tools to manage the interaction of Windows CE with the targeted hardware. However, because the platform builder is not meant to be used by application developers developing business applications, it is not covered here.

Development Tools

At the lowest level, Microsoft provides different development tools for software developers to use on the various platforms. These include the following:

- *eMbedded Visual C++ 3.0 (eVC++):* This version of Visual C++ allows for development on all Windows CE devices including embedded devices. Its primary advantages are a small footprint and speed, and it is therefore recommended when developing system-level components such as device drivers, ActiveX controls, games, and dynamic link libraries (DLLs). In addition, eVC++ is for now the only supported development environment for the SmartPhone SDK.

- *eMbedded Visual C++ 4.0:* This version can be used to develop solutions on Windows CE .NET only and is used in the same basic scenarios as mentioned earlier, namely those where performance and a small footprint are paramount in importance. Both eMbedded Visual C++ 3.0 and 4.0 produce native code (code not managed by the common language runtime).
- *eMbedded Visual Basic 3.0 (eVB):* A subset of the VB development environment that has been widely adopted for application development on the Pocket PC, this version has some limitations when compared with Visual C++ in terms of functionality and performance (all eVB code is interpreted rather than natively compiled for the device), but it is a high-productivity environment targeted at business developers. Its runtime must be installed on the device, although its larger size (more than 500K for the VB runtime and COM engine) precludes its inclusion by most OEMs on embedded devices. In all, over 150,000 developers have downloaded the free eMbedded Visual Tools, most of whom use eVB. Among its key limitations, however, are eVB's untyped nature (everything is a variant) and lack of support for object-oriented development. eVB will also no longer be updated, and so the natural path for eVB developers is to use VB .NET and the Compact Framework.
- *Microsoft Windows .NET Compact Framework with SDP for Visual Studio .NET:* The focus of this book, this version shipped with Visual Studio .NET (VS .NET) 2003. It targets Pocket PC 2000, 2002, and embedded Windows CE .NET 4.1 platforms and produces managed code using either a subset of VB .NET (VB .NET or simply VB from here on) or Visual C# .NET that is executed by the common language runtime, analogous to the desktop Framework. The Compact Framework and SDP are designed to bring mobile development to the core set of Framework developers by providing a modern and robust development environment and set of tools.

NOTE: Most code written on the Compact Framework will execute on the Pocket PC Phone Edition. However, the Compact Framework does not officially support the platform because the emulator that ships with the Compact Framework and SDP does not support the GPRS connectivity APIs, and no class libraries ship with the Compact Framework to use the APIs. To access these APIs, Compact Framework developers can call the APIs directly when the code is running on the device, using a technique called PInvoke, covered elsewhere in this book. Look for more support in a future release of the Compact Framework and SDP.

- *Microsoft ASP.NET Mobile Controls (formerly MMIT).* This also uses VS .NET, but it is a server-side technology targeting mobile devices using a Web interface and different markup languages, including HTML, WML, cHTML, and XHTML.[7] The following sections discuss Mobile Controls in greater detail.

This rich history of development support for mobile devices puts Microsoft in a strong position with managers and developers choosing a platform on which to build their mobile applications.

The Compact Framework and SDP in Context

As noted in the previous sections, the Compact Framework is targeted at a subset of the mobility platforms currently supported by Microsoft. In a nutshell, you can (and in our estimation should) consider architecting and building your applications using the Compact Framework and SDP when the following are true:

- You are developing or architecting business applications as opposed to system software.
- You are developing code that will execute on the device, as opposed to applications developed with the ASP.NET Mobile Controls.
- Your applications must run in disconnected, connected, or occasionally connected modes.
- You are targeting the Pocket PC 2000, 2002 or embedded Windows CE .NET platforms.

In addition, and as you'll learn by reading this book, the Compact Framework and SDP offer a highly productive development environment. This is primarily the case because, like its desktop cousin, the Framework, the Compact Framework offers a managed execution environment that includes a variety of services from object management to garbage collection to runtime security, freeing developers from having to perform these time-consuming, but ultimately unproductive, tasks.

Also, as we'll discuss in the next chapter, the Compact Framework includes a core set of framework classes that provide reusable code to han-

[7] Supported in Device Update 2 for VS .NET 2002 and included in the released version of VS .NET 2003.

Pocket PC on the Rise

The Pocket PC 2002 (the core platform addressed in this book) was released in the fall of 2001 in an environment in which Palm had garnered between 70% and 80% market share. Since that time, Pocket PC has made inroads in both the United States and Europe, and some analysts predict it will reach 30% of the market by 2004. However, the key focus point for this book is the market share Pocket PC already enjoys in corporations (approximately 30% and growing).

This should only increase in time as the cost of the devices drops. It also makes sense for businesses to develop on the Windows CE platform because it is an extension of the platform on which many of them already develop for the desktop and server, as IDC and Gartner have documented.

dle file I/O, XML processing, and local and remote database interaction and even to incorporate XML Web Services into your applications.

However, because the Compact Framework is a managed environment, each device that executes applications written using the Compact Framework must have the framework installed. As of this writing, there are no devices shipping that include the Compact Framework as a standard component in RAM or in ROM; however, look for future devices (Pocket PC 2003, for example) running Windows CE .NET to include the runtime.

NOTE: While Microsoft has been busy with Windows CE and the Compact Framework, other industry players have been working on smart device technology as well. The most prominent effort has been Sun Microsystems's Java 2 Micro Edition (J2ME) technology. This is Sun's version of Java aimed at devices with limited hardware resources, including PDAs, cell phones, and other consumer devices. In J2ME, each profile (analogous to a platform in the Microsoft terminology) includes a set of class libraries and a virtual machine required to support a class of device. From that perspective, J2ME is very much a direct competitor to the Compact Framework.

Many developers new to developing for mobility are at first confused by the differences between the Compact Framework and ASP.NET Mobile Controls. For those readers, and to explain why there is scant Mobile Controls coverage in this book, the following section goes into some detail as to the purpose of the Mobile Controls and the kinds of solutions for which they are appropriate.

The Role of the ASP.NET Mobile Controls

It is appropriate at this point to compare and contrast the Compact Framework with the ASP.NET Mobile Controls. The key points of similarity and difference can be summarized as follows:

KEY POINT

- The ASP.NET Mobile Controls are a server-side technology, whereas the Compact Framework is a client-side technology. This means that code written using Mobile Controls must execute on a server, not directly on the device, by producing markup language that is interpreted by the device using a browser or parser. Code written for the Compact Framework executes directly on the device using just-in-time (JIT) compilation and native execution. This concept is illustrated in Figure 1-3.
- The ASP.NET Mobile Controls are based on ASP.NET and use the ASP.NET HTTP runtime to handle and process requests via a set of ASP.NET server controls specially designed for small form factor devices. This means that Mobile Controls require an ASP.NET Web server (Internet Information Server [IIS] 5.0 or 6.0). The device

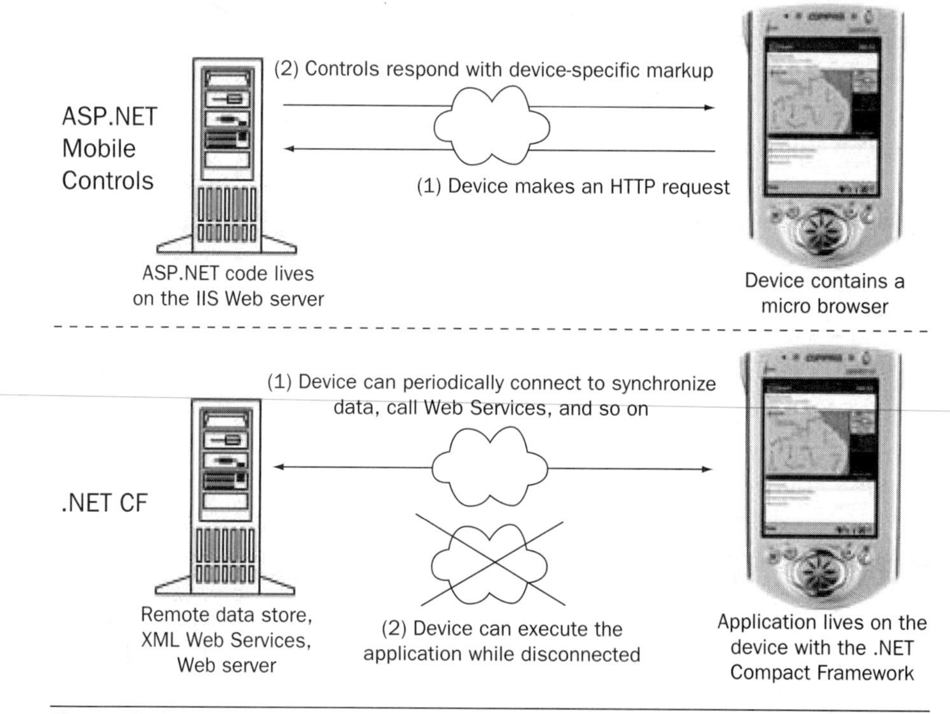

Figure 1–3 *Comparing ASP.NET Mobile Controls and the Compact Framework.*

issues requests that are processed on the Web server by the controls, producing markup language.

- Both ASP.NET Mobile Controls and the Compact Framework are based on the desktop Framework and VS .NET. In the case of Mobile Controls, developers use VS .NET to write ASP.NET code using the ASP.NET Mobile Controls. The server-side code is JIT-compiled and executed on the Web server by the common language runtime. In the case of the Compact Framework, the code is written using SDP and subsequently downloaded to the device for JIT-ing and execution using a compact version of the common language runtime.
- The ASP.NET Mobile Controls are more flexible in that they can dynamically produce various markup language, depending on the requesting device that is dynamically detected, including WML, cHTML, HTML, and XHTML. This means that Mobile Controls can support a broader range of devices (over 200 at last count), including wireless Web phones, Pocket PCs, and virtually any device that supports HTTP and HTML. The ASP.NET Mobile Controls also include the ability to add device-specific customizations, so that as new devices come online, their feature set can be identified.
- Because code written for the Compact Framework is downloaded to the device, the Compact Framework supports connected, occasionally connected, and disconnected scenarios. ASP.NET Mobile Controls support only the former because an HTTP connection is required to request an ASP.NET page that uses the Mobile Controls. Basically, this means that Mobile Controls are targeted for building Web sites for a broad variety of connected devices and, at the same time, not having to rewrite the vast majority of code to deal with differences between those devices.

Because of these basic differences in the kinds of applications, we decided it would not be appropriate to include the ASP.NET Mobile Controls in the discussions in this book. In addition, there are a number of good books on Mobile Controls (formerly MMIT) already in print and forthcoming that should provide you with adequate information to make architectural decisions. See the "Related Reading" section at the end of the chapter for a few suggestions.

The lack of coverage of Mobile Controls should not, however, be interpreted as a lack of enthusiasm for the technology. There is a large class of applications for which ASP.NET Mobile Controls are perfectly suited, for example, those developed by our own employer, Quilogy.

Quilogy Uses MMIT to Streamline Internal Processes

Quilogy (www.quilogy.com), a 300-person IT consulting and services firm headquartered in St. Charles, Missouri, had been using wireless technology since 1999, when it deployed its first internal application using custom ASP pages that produced appropriate markup languages such as WML.

Even before the release of the Mobile Internet Toolkit in 2002, Quilogy had started to rewrite its wireless applications using the beta version of the product built on top of the Framework and ASP.NET. With its release, Quilogy Wireless was born, and it now supports many of the functions of Quilogy's award-winning Intranet portal, myQ.

Quilogy Wireless on ASP.NET Mobile Controls allows Quilogy's geographically dispersed workforce (300 employees spread across 15 geographical locations) to get an integrated view of company information that is customized to their location and job role using a PIN-based logon. Some of its functionality includes viewing and responding to action items (such as approving expense reports and performing timesheet entry), accessing Microsoft Exchange e-mail and calendaring, viewing company financial information directly from the Great Plains office, viewing customer demographics, locating other employees, viewing company news, viewing recruiting and headcount information.

As a result, Quilogy's sales force, managers, and many of its consultants use Quilogy Wireless on wireless Web phones, Pocket PC phones, Palms, Blackberries, and other devices. This allows sales consultants to get the latest information on customers and management to respond to changing financial situations and employee activities in a timely fashion. For a more in-depth look at Quilogy's use of MMIT, see the article "A Quilogy MMIT Case Study" at http://atomic.quilogy.com/default.aspx?storyid=mmit1.

Quilogy also offers its customers wireless access to their own Exchange server or through its hosting service.

What's Ahead

This is a great time to be architecting solutions for mobile devices. The Compact Framework and SDP simplify that task by providing a robust execution and development environment with modern, object-oriented language support. In the next chapter we'll take apart the Compact Framework and SDP and look at their constituent parts in order to compare and contrast them with the desktop Framework. This will allow architects and developers to get a handle on the toolset and understand both its features and limitations.

Related Reading

Alesso, H. Peter, and Craig F. Smith. *The Intelligent Wireless Web*. Addison-Wesley, 2001. ISBN 0-201-73063-4.

Ferguson, Derek, and David Kurlander. *Mobile .NET*. APress, 2001. ISBN 1-893-11571-2.

McPherson, Frank. *How to Do Everything with Your Pocket PC, Second Edition*. McGraw-Hill/Osborne, 2002. ISBN 0-072-19414-6.

Wigley, Andy, and Peter Roxburgh. *Building .NET Applications for Mobile Devices*. Microsoft Press, 2002. ISBN 0-735-61532-2.

Microsoft's embedded Web site covering the operating systems Windows CE and XP Embedded, at www.microsoft.com/windows/embedded.

Microsoft mobile Web site covering the Pocket PC, Handheld PC, and SmartPhone, at www.microsoft.com/mobile.

WindowsForDevices.com. News, articles, forums focused on the Windows embedded community, at www.windowsfordevices.com.

See Chris De Herrera's Windows CE Web site for specifications, resources, and much, much more, at www.cewindows.net.

Microsoft "Mobile Devices Wireless Connectivity," white paper, at www.microsoft.com/mobile/enterprise/papers/wireless.asp.

"Pocket PC Frequently Asked Questions," at www.microsoft.com/mobile/enterprise/papers/mobilityfaq.asp.

Microsoft Mobility Web site, including case studies, white papers, and technology resources, at www.microsoft.com/Business/Mobility.

"Microsoft Navigates the Automotive Industry, Enhances Driver Experience," at www.microsoft.com/MSCorp/presspass/Press/2002/Mar02/03-04BMWpr.asp.

CHAPTER 2

Components of Mobile Development

Executive Summary

In the last two years, Microsoft has made two fundamental and important changes in its technology offerings. The first is the inclusion of XML Web Services as a cross-platform integration technology. XML Web Services relies on an XML grammar (Simple Object Access Protocol, or SOAP) to achieve platform, language, and device independence by providing a standardized, programmable way to invoke logic exposed on a Web server. Indeed, it is the hope of Microsoft and many others in the industry that XML Web Services will power a world where applications running on a variety of devices and platforms make use of a ubiquitous "Web Services fabric" to stay connected and integrate disparate silos of information, including business and personal information. Currently, there are over 80 public SOAP toolkits available for a variety of development tools and platforms.

At the same time, Microsoft was developing and did release in February 2002 Visual Studio .NET (VS .NET) and the Microsoft Windows .NET Framework. While VS .NET is a development environment, the Framework consisted of an entirely new infrastructure for executing applications and included an execution engine (EE), which is essentially the common language runtime, and a set of class libraries for developers to use. In addition, Microsoft created a new language (C#) for this environment and submitted both the specification for the EE and C# to the European Computer Manufacturers Association (ECMA), where they were later formalized. Not only do the Framework and VS .NET support cross-language development through the five languages that Microsoft supports and additional ones created by other vendors, it also enhances developer productivity by including a fully object-oriented runtime and full-featured Integrated Development Environment (IDE). Perhaps its most important feature, however, is that it includes built-in support for XML Web Services by supporting the SOAP 1.1 specification.

The .NET Compact Framework and SDP for VS .NET, shipped with Visual Studio .NET 2003, piggyback on both of these important technological shifts. They did so first by providing an EE and class library support for building applications on smart devices, allowing developers to leverage their knowledge of the Framework. They did so second by including support for calling XML Web Services, acknowledging that smart devices are an important player in the Web Services ecosystem.

The architecture of the Compact Framework allows for hardware and operating system independence by including a Platform Adaptation Layer (PAL) and Native Support Libraries (NSLs) that shield developers and allow them to write code that is portable between devices. It also includes its own rewritten EE and Garbage Collector (GC) that differ in some respects from the common language runtime and are optimized for devices. The class libraries included with the Compact Framework are roughly a subset 25% the size of the desktop Framework, with a few extensions to support the capabilities of devices.

The SDP is, of course, accessed through VS .NET 2003 and beyond and provides the project system, language support (VB and C# only in this release), graphical designers, emulators, and a debugger necessary to develop applications using the Compact Framework. In most respects the SDP leverages the existing tools found in VS .NET but also includes configuration options for the two emulators that ship with the product and the ability to debug an application both within the emulator and remotely (attached to the developer's workstation).

.NET and Smart Devices

To understand the goals of the .NET Compact Framework and SDP, it is important to review a bit of recent history, which places the goals in proper perspective.

In the summer of 2000, Microsoft began to reveal its plans for what, up to that time, had been loosely called Next Generation Windows Services, or NGWS. This amalgamation of a set of technologies and a vision for a ubiquitous Web-centric computing infrastructure coalesced under the moniker ".NET." Over the following months it came to mean two important things for technical leaders and developers: XML Web Services and the Microsoft Windows .NET Framework.

XML Web Services

In 1998, Microsoft, DevelopMentor, and UserLand Software worked together to create an XML encoding scheme that would provide direct computer-to-computer message-based interaction over the existing Internet infrastructure, including TCP/IP and HTTP as a transport. Their specification, SOAP, was submitted as an XML specification to the World Wide Web Consortium (W3C) and has been recommended and subsequently updated.[1] As a result, services built using SOAP are generally referred to as XML Web Services. At its core, therefore, a good working definition of an XML Web Service is a programmable application component accessible via standard Web protocols. Many people, Bill Gates included, have referred to the advent of XML Web Services as the "third wave" of the Internet, following the connectivity and presentation waves that preceded it.

Obviously, the importance and promise of XML Web Services are its use of Internet standards and independence from the technologies that characterized the "component wars" of the 1990s, which were often reduced to battles between Microsoft's Distributed Component Object Model (DCOM)[2] and the Common Object Request Broker Architecture's (CORBA)[3] Internet Inter-ORB Protocol (IIOP). This independence means that XML Web Services tear down three primary barriers to interoperation and integration as shown in Figure 2-1.

- *Platform:* Because SOAP has nothing to say about the technology used to create and consume SOAP messages and because XML parsers exists for all platforms, various vendors can implement toolkits and add-ons for their platforms to create and consume SOAP messages. In fact, there are over 80 SOAP toolkits available today. Removing the platform barrier, for example, goes a long way toward allowing software running on some flavor of Unix to communicate seamlessly with software written on Windows.[4]
- *Language:* Because SOAP toolkits are available for multiple platforms, a variety of languages can be used on those platforms to create

[1] The current SOAP specification is 1.1 and can be found at www.w3c.org/TR/2000/NOTE-SOAP-20000508.

[2] DCOM is based on the original COM specification from Microsoft.

[3] CORBA is an architecture for distributed objects created by the industry consortium Object Management Group (OMG)

[4] To speed standardized adoption of SOAP, Microsoft and IBM created the Web Services Interoperability Organization (WSI). See www.ws-i.org for more information.

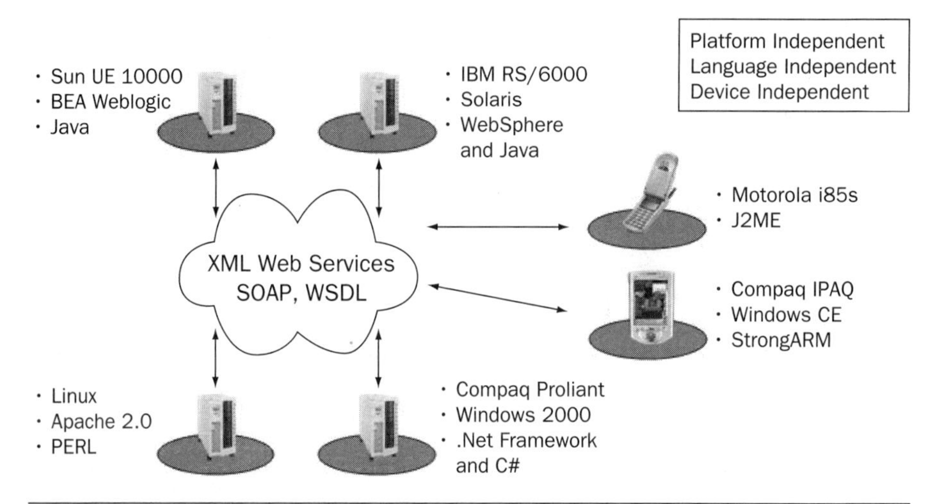

Figure 2–1 *Web Services Interoperability.* This diagram highlights the platform-, language-, and device-independent nature of XML Web Services.

and consume SOAP. For example, IBM in its WebSphere product allows XML Web Services to be created using Java, while Microsoft includes XML Web Services creation in Visual Studio .NET (VS .NET) and Visual Studio 6.0 using VB .NET, and VB 6.0.

- *Device:* Most important for our purposes in this book, XML Web Services break down the barriers between devices. This is the case because SOAP messages simply consist of XML and are, therefore, independent of the device (from a desktop PC to a Pocket PC to an appliance to a gaming console) on which they are consumed and their results displayed.

Indeed, Microsoft and others envision a world where applications running on a variety of devices make use of a Web Services fabric to stay highly connected and bring together disparate silos of information, including business and personal information.

One of the reasons you're reading this book is that Microsoft and other leading industry organizations are betting that mobile devices will play a key role in the XML Web Services ecosystem.

Windows .NET Framework

The second important concept that grew out of .NET and was released along with VS .NET 2002 in February of 2002 was obviously the Framework.

Although we assume most readers are familiar with the .NET development platform, a brief introduction for those who aren't is provided here. For an in-depth look at the common language runtime, see the books by Don Box, Dan Fox, and Jeffery Richter in the "Related Reading" section at the end of the chapter.

Simply put, the Framework consists of an EE, or virtual machine, referred to as the common language runtime, and a set of class libraries that provide the low-level interaction with the underlying operating system—often called the Base Class Libraries (BCL). Also provided are higher-level classes that expose programming models for developing specific types of applications and utilizing system services—often called the Services Framework.[5] Both of these components were released along with VS .NET 2002, which provided the IDE and language support for creating applications that utilize the Framework. However, in the initial release, applications written for the framework (referred to as managed code) were restricted to desktop PCs and servers running Windows 98, ME, NT, XP, and 2000. As you can imagine, the class libraries and the runtime engine must be present on any machine running managed code.[6]

In this scheme, the common language runtime includes a host of runtime features including a class loader, thread support, exception manager, security engine, GC, code manager, and type checker. In turn, all managed code is first compiled to a machine-independent intermediate language called Microsoft Intermediate Language (MSIL) and subsequently compiled to native instructions for execution in a JIT manner, as the common language runtime's class loader loads code at runtime.

When the code is compiled to MSIL by the developer using VS .NET, it is stored in a portable executable (PE) file called a module. The module contains the MSIL instructions, in addition to metadata that describes the types (classes, interfaces, enumerated types, and so on) in the code the developer has written, along with the dependencies on other types. The common language runtime and other tools in VS .NET rely heavily on this metadata to make sure that the appropriate code is loaded and to assist in enabling features such as IntelliSense and debugging in VS .NET. A module can then be incorporated into, or exist independently as, an assembly. An assembly is the fundamental unit of packaging, deployment, security, and versioning in .NET and contains a manifest (embedded in one of the modules or in its

[5] These include ASP.NET, ADO.NET, and Windows Forms, among others.

[6] Windows .NET Server was the first operating system to ship with a version of the Framework (v1.1) preinstalled. Other avenues include a redistributable package available for developers to include with their code or the Windows Update Service.

own PE file) that describes the version, an optional public key token used for uniquely identifying this assembly from all others, and a list of dependent assemblies and files.

It should be noted, however, that the rich nature of MSIL makes assemblies a prime target for reverse engineering. To avoid this, a number of third parties provide obfuscation products that work by renaming the symbols in the MSIL to make it more difficult for disassemblers to understand the code. The community version of PreEmptive Solutions Dotfuscator is included with VS .NET 2003 and can be used to obfuscate both desktop Framework and Compact Framework assemblies.

The object-oriented class libraries that contain roughly 6,000 types are organized into hierarchical namespaces (for example `System`, `System.Data`, `System.Windows.Forms`, and so on) and shipped in assemblies that are installed with the framework. These integrated class libraries are a boon for Windows developers because they unify the previously used programming models into a single framework that includes a rapid application development (RAD) forms package, full object-orientation, XML and data support, and full support for writing Web applications. They also expose new concepts to developers, including built-in support for creating and consuming XML Web Services. These programming models are then fully exposed through project templates in VS .NET.

Finally, it's important to note that the Framework is at its core language agnostic. Microsoft now ships five language compilers (VB, C#, J#, JScript, and Visual C++ with Managed Extensions) that produce MSIL and, thus, support building managed code. The rich metadata contained in assemblies and the use of MSIL serve to abstract language differences that in the past raised barriers when interoperating between languages. The picture that emerges can be summarized in Figure 2-2. Figure 2-2 highlights the architecture of the Framework and its role in application development. Note that VS .NET templates, composed of developer source code, and references to the Services Framework and BCL are then compiled into assemblies before being executed by the common language runtime.

To summarize, the Framework and Visual Studio .NET provide several key benefits to developers and their organizations:

- *Developer productivity:* The power of the unified programming model exposed in the Framework enables developers to build applications more easily and removes the glass ceiling felt by VB developers. In addition, VS .NET, with its plethora of features, including IntelliSense, code outlining, graphical designers, commenting, and

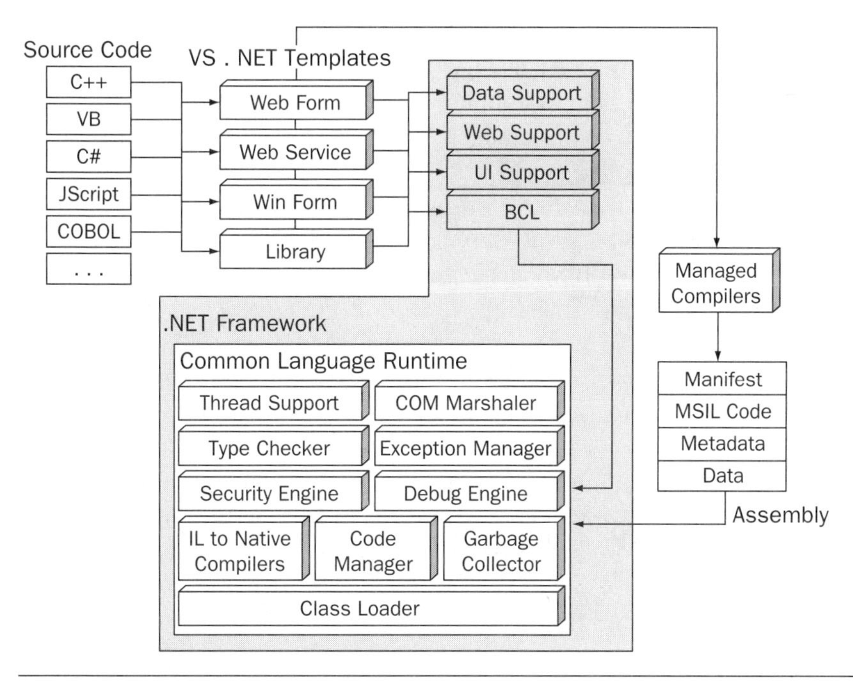

Figure 2–2 *The Microsoft Windows .NET Framework.*

integrated debugging just to name a few, enables developers to build applications more quickly.

- *Modern languages:* As a corollary to developer productivity, the inclusion of modern object-oriented languages like C# and VB .NET allows developers to be more productive by using features such as inheritance and polymorphism. This benefit should not be underestimated, especially for developers coming from the procedural and weakly typed world of eVB.
- *Language choice:* Although the .NET Framework may in the future be ported to CPU architectures other than x86,[7] it can more rightly be thought of as a multilanguage, rather than a multiplatform, environment. In fact, more than 25 languages from ISVs and academic

[7] For example, there is an open source project called "Mono" that is porting the Framework to Linux, and Microsoft has released a FreeBSD implementation for academic use called "Rotor." For more information see www.go-mono.com. In addition, both C# and the Common Language Infrastructure (CLI) specification, which describes the runtime engine, were submitted and have been approved by ECMA as specifications ECMA-334 and ECMA-335, respectively. The common language runtime is a superset of the CLI.

institutions are either released or in the works.[8] This allows organizations to leverage their existing investment in these languages, rather than having to retrain their development staff. However, the abstraction provided by managed code also made it easier to create the .NET Compact Framework for devices, as we'll see later in this chapter.

- *XML Web Services integration:* By integrating support for XML, SOAP 1.1, Web Services Description Language (WSDL[9]), and other XML Web Services specifications, the Framework allows developers to create a new class of applications that are not only positioned to take advantage of the future Web Services fabric, but also provide a great way to promote intraorganization integration of line-of-business and legacy applications.

As you can see, the myriad benefits of XML Web Services, coupled with the Framework, provide a compelling reason for development organizations to adopt .NET technology. And, because of its success, Microsoft has now extended these benefits to smart devices by releasing the .NET Compact Framework and SDP.

Goals of the Compact Framework and SDP

As mentioned in the previous section, the release of VS .NET 2003 has served to bring the benefits of XML Web Services and the Framework to smart devices. As a result, the goals of the Compact Framework and SDP can be enumerated as follows:

- *Portable (and small) subset of the Windows .NET Framework targeting multiple platforms:* This includes the creation of a runtime engine, class libraries, and compilers for VB and C# that together are referred to as the .NET Compact Framework. As you'll see in the next section, this framework is not a simple port of the desktop version, but a complete rewrite designed to execute managed code on multiple CPU architectures and operating systems through an abstraction layer known as the PAL. It is important to note that the

---

[8] For example, Fujitsu COBOL for the Framework. For more information, see www.adtools.com/info/whitepaper/net.html.

[9] Pronounced "wizz-dull." WSDL is an XML grammar used to describe the operations exposed by a Web Service.

Compact Framework was also designed to be a subset of the desktop version of the desktop Framework. As a result, roughly 25% of the desktop types are represented in the Compact Framework.

- *Leveraging of VS .NET:* Because VS .NET already provides a high-productivity development environment, it is only natural that this environment should be utilized in developing mobile applications as well. To that end, SDP for VS .NET 2003 provides the project templates, emulators, debugger, and device integration to use the same IDE for desktop as for mobile development. SDP will be explored in detail in the second half of this chapter.
- *True emulation:* One of the requirements for developing robust (i.e., well-debugged) mobile applications is that they run as expected when installed on the device. To that end VS .NET includes emulators for Pocket PC and Windows CE that execute exactly the same operating system binaries, EE, and class libraries as those installed on the device. In addition, the emulator supports localized packages for developing global applications, as will be explored in Chapter 8. In this way, developers can be assured that the code they write and test with the emulator will execute correctly when deployed to the device.
- *Enabling of Web Services on devices:* As mentioned previously, because XML Web Services are by definition device-independent, they can be used from a variety of devices. By building support for consuming Web Services directly into the Compact Framework, developers can easily call Web Services, as will be discussed in Chapter 4.

Taken together, the realization of these goals means that the benefits of XML Web Services and the desktop Framework discussed in the previous section can be realized on devices as well.

The remainder of this chapter will focus on the architecture and features of the Compact Framework and SDP.

The .NET Compact Framework

In this section, we'll explore the architecture, core functionality, and UI support in the Compact Framework and compare it with the desktop Framework. In this way, developers and technical managers can quickly get a feel for the technology involved in developing mobile applications using the Compact Framework.

Architecture

You may recall that one of the design goals of the Compact Framework was to create a "portable (and small) subset of the desktop Framework, targeting multiple platforms." To support this goal Microsoft created the architecture shown in Figure 2-3. In this section we'll walk through the components of that architecture from the bottom up to describe how each contributes to this goal.

The size of the Compact Framework installed on the device varies from device to device. Generally, it ranges from 1.7MB to 2.6MB and can be installed in RAM, ROM, or FlashROM. As you might expect, the initial release is installed in RAM so that it is immediately available to all devices. However, in the future, expect OEMs to offer FlashROM upgrades that include the Compact Framework for devices like the Pocket PC 2002 (where FlashROM is already present). Future devices (i.e., Pocket PC 2003 devices) will likely ship with the Compact Framework in ROM.

Host Operating System

Obviously, at the lowest level, the code written for the Compact Framework must be executed on a host operating system such as Windows CE. At this time, the Compact Framework will run on Windows CE 3.0 and Windows CE .NET 4.1, although as we'll see, the architecture in Figure 2-3, particularly through the inclusion of the PAL and the NSLs, lends itself to portability to other host operating systems as well.

PAL

The PAL is the primary component that makes platform portability possible. Essentially, the PAL contains a variety of subsystems that expose the functionality of the underlying operating system and hardware in a consis-

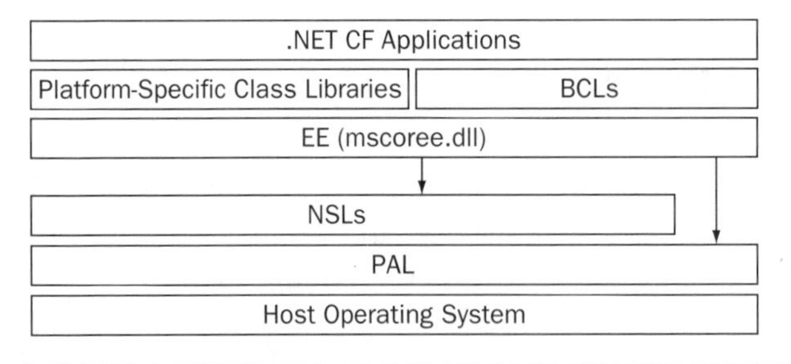

Figure 2–3 *The Compact Framework Architecture.*

tent set of APIs to the NSL and EE, as shown in Figure 2-3. For example, the PAL includes interfaces for device drivers, a system memory manager, interrupts and timers, multimedia, and I/O ports, among others. All of these subsystems must be fully implemented on the target device.

As a result, in order to port the Compact Framework between devices, OEMs must rewrite the PAL to make native calls on the target operating system and to the hardware. Of course, depending on the features of the device and what its native operating system supports, the functions of the PAL may or may not map to the operating system in a straightforward manner. For example, the PAL was designed with Windows CE and the Pocket PC in mind, and so, many of its APIs simply map directly to APIs exposed by Windows CE.

In summary, you can think of the PAL as the equivalent of a device driver used by the Compact Framework that abstracts and drives the underlying operating system and its hardware.

Native Support Libraries

Because not all devices support the same set of services, the Compact Framework also includes a set of NSLs that implement features that the Compact Framework requires, including file system operations, heap management, globalization, cryptography, and graphical user interface (GUI) manipulation.

These services then make calls into the PAL to perform their operations and are in turn called by the EE, as shown in Figure 2-3. A typical example is the GUI support implemented as an NSL that is then exposed by the classes in the `System.Windows.Forms` namespace in the Compact Framework class library.[10] As you might expect, the NSLs can also be called by other native code running on the device, thereby substantially increasing the feature set available to other unmanaged applications as well.

The addition of NSLs levels the playing field for devices that do not support these core features, making the Compact Framework capable of running on a wide variety of devices.

Because the NSLs make use of the PAL, OEMs do not have to port the code to implement them as they do with the PAL. The NSLs can simply be compiled for the target platform.

---

[10] This NSL is implemented by the file Netcfagl1_0.dll installed on the device. On Windows CE, this Advanced Graphics Library interfaces with the Graphics, Windowing, and Event Subsystem (GWES), as shown in Figure 1-2.

EE

The EE in Figure 2-3 provides essentially the same set of services that the common language runtime does for desktop and server applications shown in Figure 2-2 by managing the execution of a .NET application. However, because it performs these functions in an environment where resources are scarce (on devices with less memory and a slower CPU), the EE was designed from the ground up with these constraints in mind and, as a result, performs some of them differently. Even so, the core technology, like the desktop Framework, still conforms to the ECMA-335 specification. The EE was written in C and is implemented in two DLLs, `Mscoree.dll` (the stub) and `Mscoree1_0.dll` (the bulk of the EE), which, like the NSLs, are compiled for the target platform per CPU and operating system.[11]

To get a better understanding of how the EE does its work, the following list explicates some of the core functionality in the order it is encountered during the execution of a managed application:

- *Class Loader:* As with the desktop Framework, code executed by the Compact Framework must have been previously compiled into MSIL instructions and placed in an assembly (a PE file) on the device. The compilation occurs on the developer's PC using SDP, as explained later in this chapter. As the name implies, the job of the Class Loader is to locate and load the assemblies required to execute an application. However, before the Class Loader can do its work, the application must be activated at the operating system level, which occurs when the Compact Framework application is executed by the host operating system.[12] At that time, a process is created, and `Mscoree.dll` (and subsequently `Mscoree1_0.dll`) is loaded by the operating system[13] into the process. At this point an Application Domain is created, and the EE takes over execution of the application within the domain running in the operating system process.[14] As with the desktop Framework, Application Domains serve as a means to isolate Compact Framework applications running within the same

[11] Typically, the EE ranges in size from 400K to 500K depending on the operating system and CPU architecture.

[12] For example, the Windows CE PE Loader.

[13] For example, using the Windows CE `LoadLibrary` API.

[14] The APIs required to create and manage Application Domains within a custom host process are not documented in the initial release of the Compact Framework. It is also not possible to load assemblies into a domain-neutral code area for use by multiple Application Domains.

process and can therefore be thought of as "lightweight processes." Once the EE has been invoked, the Class Loader can then do its job by loading the set of assemblies, with the required versions, necessary to execute the application. It does this by inspecting the metadata in the assembly that includes the information about dependent assemblies. The list of required assemblies can (and often does) include both custom assemblies that developers create and assemblies that ship with the Compact Framework, such as `System.Windows.Forms`.[15] The Compact Framework Class Loader uses a simpler scheme for binding than does the desktop Framework. In short, the Class Loader looks at the major and minor version number (of the four-part naming scheme) of the referenced assembly and will load it as long as they are the same as the version the calling assembly was compiled with. The Class Loader also supports side-by-side execution, which means that a Compact Framework application always runs on the version of the EE with which it was compiled.

- *Type Checker:* After the Class Loader has loaded the required assemblies, the Type Checker is invoked to determine if the MSIL code is safe to execute. In this way, the Compact Framework provides the same verifiably type-safe execution as the desktop Framework, for example, by making sure that there are no uninitialized variables, that parameters match their types, that there are no unsafe casts, that the array indexes are within bounds, and that pointers are not out of bounds.
- *JIT compiler:* Once the Type Checker has verified the code and completed successfully, the MSIL code can be JIT-compiled to native instructions on the CPU. And, as with the desktop Framework, the compilation occurs on a method-by-method basis as each method is invoked. However, the Compact Framework JIT compiler must be especially sensitive to the resource constraints of the device and so uses a code-pitching technique to free blocks of memory when resources are low. This works by marking sections of JIT-compiled code that were recently executed and then allowing the least recently executed blocks to be reclaimed in a process similar to that used by a GC. As with most things, this too is a trade-off because MSIL code must be recompiled if it is subsequently executed. However, using this technique typically ensures that the core working set of the application stays natively compiled in memory. It should be noted

[15] The Compact Framework, however, does not support multifile assemblies as the desktop Framework does.

that the Compact Framework does not support compiling an entire application to native code at install time using the native code generation (`Ngen.exe`) command-line utility as the desktop Framework does.

- *Thread support:* As a Compact Framework application runs, it can gain access to underlying operating system threads through the `Thread` class in the `System.Threading` namespace. This allows Compact Framework developers to create applications that appear more responsive by offloading work (for example, a call to an XML Web Service) to background threads, while waiting for stylus input from the user.[16] It should be noted that the Compact Framework was designed to coexist peacefully with the host operating system and so relies on native operating system threads and synchronization primitives. As a result, operating system scheduling priorities also apply to Compact Framework applications, and threads produced by the EE can coexist with native threads in the same process. In addition, the Compact Framework includes a thread pool (`System.Threading.ThreadPool`) that allows a developer to queue a method for execution on one of a number of background worker threads controlled by the EE. When a thread in the pool is free, the method will execute and, when finished, can notify the main thread through a callback.
- *Exception handling:* During the execution of an application, unforeseen events sometimes transpire. To handle these gracefully, the Compact Framework supports structured exception handling (SEH), as does the desktop Framework. This allows developers to use `Try-Catch` semantics in their code and to test for specific types of exceptions thrown by the application. And, as with its desktop cousin, the EE of the Compact Framework is optimized for the nonexceptional case, and so throwing exceptions should be reserved for true exceptions and not simply to signal a normal occurrence. The key difference in exception handling in the Compact Framework is that the error strings are actually stored in a separate assembly, `System.SR.dll`. This is due to the resource constraints of devices and allows the developer optionally to install this assembly on the device. If present, the EE will load it and display the appropriate message, and, if not, a default message will be displayed.
- *GC:* One of the most discussed features of the common language runtime is the GC, which is responsible for managing memory by

[16] The Application Domain, in which a multithreaded Compact Framework application runs, will exist until all of the created threads have exited.

collecting and deallocating objects that are no longer used. As you might expect, the design of the GC is especially important in the constrained environment of a mobile device, and, for that reason, it differs from the GC implemented in the desktop Framework. The GC in the Compact Framework consists of an *allocator* and a *collector*. The allocator is responsible for managing the object pools that provide storage for the instance data associated with an object, while the collector implements the GC algorithm. At a high level, the collector runs on a background thread when resources are low and, while working, freezes all other active threads at a safe point. It then finds all reachable objects by traversing the various thread call stacks and global variables and marks them. The collector then frees all the unmarked objects and executes their finalizers.[17] Finally, the object pools are compacted, which returns free memory to the global heap. This approach is referred to as a "mark-and-sweep approach" and does not use the concepts of generations or implement a finalization queue, as does the more complex GC of the desktop Framework.

In addition to the features discussed here, the EE also provides other services that will be discussed in more detail in the following section, including exception handling, native code interoperation, and debugging.

Class Libraries

In order to create a robust programming environment for devices, the Compact Framework ships with a set of class libraries in assemblies organized into hierarchical namespaces similar to those found in the desktop Framework described previously and shown in Figure 2-3. However, there are four major differences between the class libraries shipped with the desktop Framework and those included in the Compact Framework.

KEY POINT

First, the Compact Framework libraries can rightly be thought of as a subset of the desktop libraries and, in fact, include just over 1,700 types, or roughly 25% of the desktop Framework. The diagram shown in Figure 2-4 highlights the namespaces supported in the Compact Framework as a subset of those in the desktop Framework, where there is overlap. As a result, developers familiar with the desktop Framework should not expect all the

[17] Finalizers are the destructors associated with an instance of a class. Destructors are often used explicitly to free resources but are not required.

functionality they are accustomed to. The most important additional omissions are listed here:

- *ASP.NET:* Because the Compact Framework is designed to support applications that execute on the device, it does not include any support for building Web pages hosted on a Web server running on the device. This means that the classes of the `System.Web` namespace familiar to ASP.NET developers are not found in the Compact Framework. To write Web applications that can be accessed by a mobile device, use the ASP.NET Mobile Controls as discussed in Chapter 1.
- *COM Interop:* Because the Windows CE operating system and the eVC++ tool support creating COM components and ActiveX controls, it would be nice if the Compact Framework supported the same COM Interop functionality (complete with COM callable wrappers and interop assemblies) as does the desktop Framework. Unfortunately, COM Interop did not make it into the initial release of the Compact Framework. However, it is possible to create a DLL wrapper for a COM component using eVC++ and then to call the wrapper using the Platform Invoke (PInvoke) feature of the Compact Framework, which allows native APIs to be called. Examples of using PInvoke can be found throughout this book, but especially in Chapter 11.
- *OleDb access:* The Compact Framework omits the `System.Data.OleDb` namespace and so does not support the ability to make calls directly to a database using the OleDb .NET Data Provider. However, the remote data access (RDA) features of SQL Server CE do support pulling data down from a SQL Server that can act as a repository for data from other data sources, as discussed in Chapter 7.
- *Generic serialization:* The desktop Framework supports binary and SOAP serialization of any object through the use of the `Serializable` attribute, the `ISerializable` interface, and the `XmlSerializer` class in the `System.Xml.Serialization` namespace. This functionality is not supported in the Compact Framework. However, the Compact Framework does support serializing objects to XML for use in XML Web Services and serializing `DataSet` objects to XML as discussed in Chapter 3.
- *Asynchronous delegates:* Delegates in both the desktop Framework and Compact Framework can be thought of as object-oriented function pointers. They are used to encapsulate the signature and address of a method to invoke at runtime. While delegates can be called synchronously, they cannot be invoked asynchronously and passed a call-

back method in the Compact Framework. However, it should be noted that asynchronous operations are supported for some of the networking functionality found in the `System.Net` namespace and when calling XML Web Services described in Chapter 4. In other cases, direct manipulation of threads or the use of a thread pool is required as described in Chapter 3.

- *Application configuration files:* The desktop Framework includes a `ConfigurationSettings` class in the `System.Configuration` namespace. This class is used to read application settings from an XML file associated with the application and called `appname.exe.config`. The Compact Framework does not support this class, but developers can write their own using the classes in the `System.Xml` namespace discussed in Chapter 3. An example class of this type can be found in the book by Wigley and Wheelright referenced in the "Related Reading" section at the end of the chapter.
- *.NET remoting:* In the desktop Framework, it is possible to create applications that communicate with each other across application domains using classes in the `System.Runtime.Remoting` namespace. This technique allows for data and objects serialized to SOAP or a binary format to be transmitted using TCP or HTTP.[18] This functionality is not supported (in part because generic serialization is not supported) in the Compact Framework, where, instead, XML Web Services and the Infrared Data Association (IrDA) protocol can be used, as discussed in Chapter 4.
- *Reflection emit:* Although the Compact Framework does support runtime type inspection using the `System.Reflection` namespace,[19] it does not support the ability to emit dynamically created MSIL into an assembly for execution.
- *Printing:* Although the Compact Framework does support graphics and drawing through a subset of the GDI+ functionality of the desktop Framework, it does not support printing through the `System.Drawing.Printing` namespace.[20]

---

[18] See Chapter 8 of *Building Distributed Applications with Visual Basic .NET* for an overview of .NET Remoting.

[19] For example, to create objects dynamically at runtime using the `Activator.CreateInstance` method.

[20] The most popular third-party printing software is the PrinterCE SDK from Field Software Products (www.fieldsoftware.com). Look for a Compact Framework version of their SDK in the near future.

- *XPath/XSLT:* Support for XML is included in the Compact Framework and allows developers to read and write XML documents using the `XmlDocument`, `XmlReader`, and `XmlWriter` classes, as discussed in Chapter 3. However, it does not support executing XPath queries or performing XML Stylesheet Language (XSL) transformations.
- *Server-side programming models:* As you would expect, in addition to those shown in Figure 2-4, the Compact Framework also does not support the server-side programming models, including `System.EnterpriseServices` (COM+),[21] `System.Management` (Windows Management Instrumentation, or WMI),[22] and `System.Messaging` (Microsoft Message Queue Server, or MSMQ).[23]

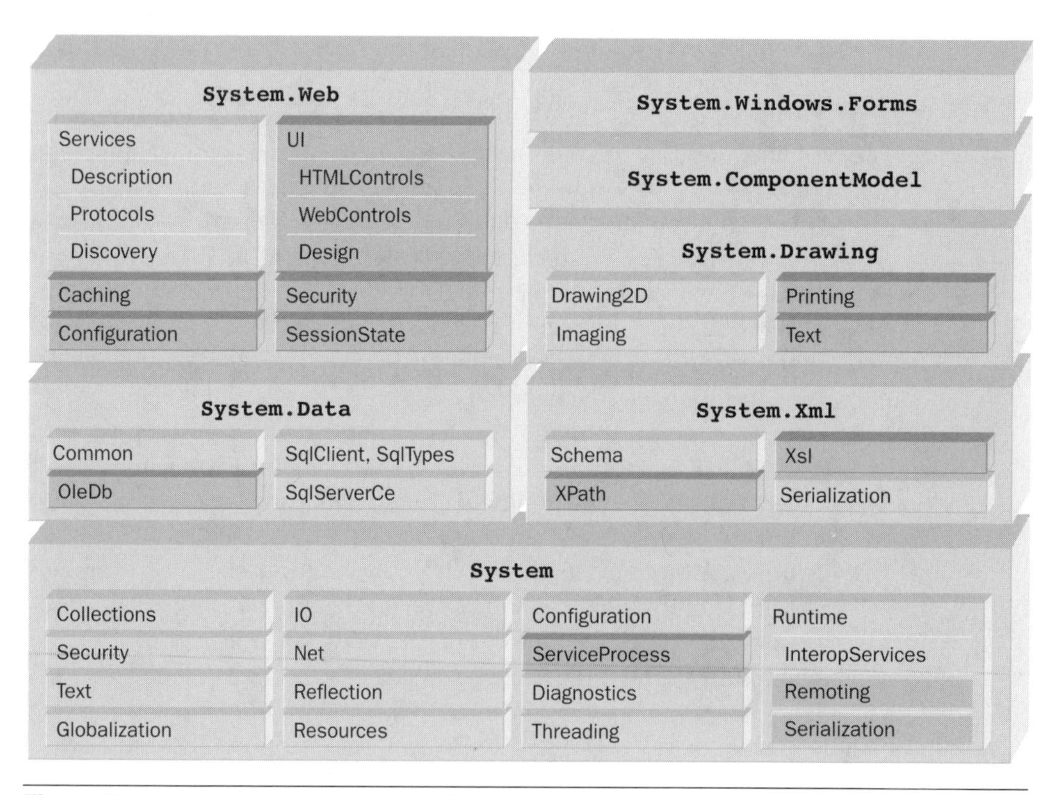

Figure 2–4 *The Compact Framework Class Libraries.*

[21] Or Component Services that enable .NET components to access services such as distributed transactions, object pooling, and loosely coupled events.

[22] Used to monitor services on a Windows machine.

[23] Used to create asynchronous message-based applications.

- *Multimodule assemblies:* The desktop Framework supports the ability to deploy an assembly as a collection of files. This is useful for creating assemblies authored with multiple languages. This feature is not supported in the Compact Framework where a single file (.exe or .dll) represents the entire assembly.

The second major difference between the desktop class libraries and those included in the Compact Framework is how they are factored into assemblies. Simply put, in the Compact Framework the 14 assemblies that comprise the class libraries are more granular than those found in the desktop Framework. For example, in the desktop Framework, the classes of the `System.Data.SqlClient` namespace used to access SQL Server are included in the `System.Data.dll` assembly, whereas in the Compact Framework they are factored into their own assembly. In this way, the Compact Framework can support a smaller footprint if applications installed on the device do not require some of the Compact Framework class libraries.

The third major difference is that the Compact Framework supports two additional namespaces (each shipped in its own assembly) that expose functionality particular to smart devices, as shown in Table 2-1.

In addition, support for the IrDA protocol has been included in the Compact Framework and exposed in the six classes found in the `System.Net.Sockets` and the `System.Net` namespaces, as covered in Chapter 4.

Finally, the Compact Framework supports a subset of the core types supported in the desktop Framework. These are often referred to as the Common Type System (CTS) types because they are the foundational types. Table 2-2 presents the CTS types found in the desktop Framework

Table 2–1 Additional Compact Framework Namespaces

Namespace	Use
`Microsoft.WindowsCE.Forms`	Contains the `InputPanel`, `MessageWindow`, and `Message` classes. The `InputPanel` class is used to control the soft input panel (SIP) on a Pocket PC, the primary means of entering data. The `MessageWindow` and `Message` classes are used primarily to communicate with unmanaged (or managed) applications using standard Windows messaging.
`System.Data.SqlServerCE`	Contains approximately 30 classes used to communicate with SQL Server CE if installed on the device.

Table 2–2 Types Supported in Desktop and Compact Framework

Type	Description
`System.Object`	Root object of the CTS hierarchy from which all types derive
`System.Delegate`	Used to store the address of a method to invoke; basis of event handling; asynch is not supported
`System.Array`	Base type for all arrays
`System.Boolean`	True or false, 8-bit unsigned
`System.Byte`	Unsigned 8-bit integer
`System.Char`	A Unicode 16-bit character
`System.DateTime`	Stores date and time information from 1/1/0100 12:00:00AM in 100-ns tick intervals
`System.Decimal`	Represents positive and negative values with up to 28 significant digits
`System.Double`	IEEE 64-bit float
`System.Int16`	Signed 16-bit integer
`System.Int32`	Signed 32-bit integer
`System.Int64`	Signed 64-bit integer
`System.IntPtr`	Signed integer, native size
`System.SByte`	Signed 8-bit integer
`System.Single`	IEEE 32-bit float
`System.TimeSpan`	Represents a time interval
`System.UInt16`	Unsigned 16-bit integer
`System.UInt32`	Unsigned 32-bit integer
`System.UInt64`	Unsigned 64-bit integer

and their support in the Compact Framework. Note that in each case, some of the overloaded methods and properties found in the desktop Framework are not supported in the Compact Framework.

Developers familiar with eVB will no doubt note the vast difference between the strongly typed Compact Framework environment and eVB, where all variables are of the type `Variant`. This not only makes source code easier to read and avoids runtime errors, it also saves memory because in eVB each variable consumes a minimum of 16 bytes, whereas the types in the Compact Framework consume fewer; for example, `System.Int32` is four bytes.

In all, the Compact Framework class libraries provide a wealth of functionality that allows developers to create robust applications for devices.

Portability

All development teams would like to be able to leverage the code they write by reusing it in as many scenarios as possible, and the developers of Compact Framework code are no exception. There are four key scenarios where portability must be addressed: device to device, desktop to device; device to desktop, and eMbedded Visual Tools to Compact Framework.

Device to Device

Although devices targeted for the Compact Framework span a variety of hardware manufacturers and processors, the architecture illustrated in Figure 2-3 allows Compact Framework applications to be moved between devices without recompilation. This is the case for two reasons. First, the MSIL code placed in the assembly by the compiler is machine-independent, allowing the JIT compiler of the EE to compile to native instructions for execution. In addition, the system assemblies contain the same set of classes, are factored identically, and are versioned identically on all platforms. This allows your development team to create a single binary to support multiple devices running on multiple CPUs (x86, SH3, ARM, MIPS), all of which are loaded with the Compact Framework.

In this scenario the caveat is that Compact Framework applications also support several platforms (Pocket PC 2000, 2002, Windows CE .NET 4.1) and any particular application may rely on platform-specific features, for example, the `InputPanel` class to control the SIP on Pocket PC. In these cases, the application would need to be modified to remove the unsupported features and recompiled before executing it on another platform. Not doing so may result in exceptions being thrown or unpredictable behavior. For example, if a Compact Framework application targeted for Windows CE .NET 4.1 attempts to display the SIP using the `InputPanel` class, the EE throws a `NotSupportedException`. To work around these issues, it is possible to determine programmatically the platform using a native API call as discussed in Chapter 11.

Desktop to Device

Porting an application from the desktop Framework to the Compact Framework, however, is not so simple. This is the case because of the differences between the class libraries in the Compact Framework as discussed in the previous section. Fortunately, because the design philosophy of the Compact Framework team at Microsoft was to provide both familiarity and full compatibility in the developer experience, the core programming model, language support, and file formats and protocols, porting an application from the desktop Framework to the Compact Framework is largely a matter of removing functionality not supported because of the subset nature of the Compact Framework and recompiling the application in an SDP in VS .NET. You or your team will of course also need to redesign the UI to work within the smaller screen size and other constraints of the device, as well as adding any platform-specific features.

Even in the simplest case, if an assembly is created in the desktop Framework and then referenced in an SDP in VS .NET, both a warning dialog and compiler errors will result, indicating that the `mscorlib` assembly referenced by the desktop Framework assembly differs from that referenced by the SDP. For this reason, it is recommended that all desktop Framework assemblies for use on the Compact Framework first be recompiled.

Interestingly, it is possible to load a desktop Framework assembly on the Compact Framework using the `Assembly` class of the `System.Reflection` namespace. However, because the Compact Framework does not support late binding, invoking the methods requires more runtime type inspection (using the `Type` and `MethodInfo` classes) and is therefore unwieldy at best.

Device to Desktop

This scenario has much in common with the previous one. Although the Compact Framework uses the same standard PE file format, header, and metadata as that used by the desktop Framework, applications created for the Compact Framework will be designed for the constraints of the device and will use platform-specific assemblies (`Microsoft.WindowsCE.Forms`) and functionality (IrDA). For this reason, most Compact Framework code will need to be modified and recompiled in a desktop Framework project for execution on the desktop.

KEY POINT

However, it is possible to create assemblies for the Compact Framework that the desktop Framework will load and execute without recompilation. This works because the desktop Framework supports assembly retargeting for the assemblies that have counterparts between the two systems.[24] In other words, even though a Compact Framework application references the Compact Framework–specific `System.Data.dll` assembly (that has its own strong name and is a subset of the desktop Framework version), at

[24] This works through the inclusion of the `retargetable` modifier added to the MSIL code by the compiler.

runtime the common language runtime can retarget this reference to the full desktop Framework version and bind accordingly. In future versions of the desktop Framework, look for the addition of a retargeting publisher policy that can be applied to allow administrators to control how and when retargeting is applied.

Of course, because of the UI, memory, and other constraints of devices, this sort of binary compatibility will likely be useful only for developing custom code libraries (business logic, sorting, searching, data access, string, and file manipulation, for example).

eMbedded Visual Tools to Compact Framework

As mentioned in Chapter 1, before the availability of the Compact Framework, developers creating applications for smart devices used eVC++ and eVB that together are referred to as the eMbedded Visual Tools. However, because the Compact Framework uses an entirely new EE, class libraries, along with a new IDE and language syntax, porting an application written in eVB will require a substantial rewrite. And unlike in the desktop Framework, there is no tool available in VS .NET to assist in the upgrade process. For this reason, development teams will likely port only eVB applications to Compact Framework when adding significant additional functionality, for example, by adding support for XML Web Services.

SDP

The second half of this chapter deals with the features and functionality of SDP. Simply put, SDP includes the components that allow Compact Framework applications to be developed using VS .NET. The purpose of this section is to bring architects and developers quickly up to speed on how SDP leverages VS .NET by integrating with the development environment and languages, on the UI tools available, how the Compact Framework provides true emulation, how the debugging process works, and how other tools can assist in the development process.

Project System

In VS .NET 2003 to create an SDP, a developer simply needs to select the Smart Device Application icon in the New Project dialog. When selected, the Smart Device Application Wizard will be invoked, as shown in Figure 2-5.

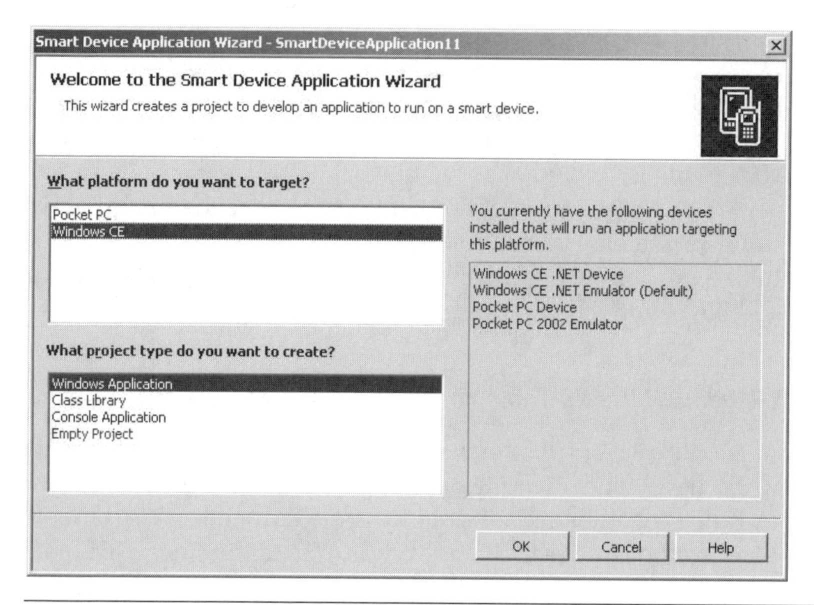

Figure 2–5 *Creating an SDP.*

As you can see in Figure 2-5, SDP supports two platform types, Pocket PC and Windows CE .NET; and for each one, a set of project types follows:

- *Windows application:* This creates a form-based application using the `Windows.Forms` assembly. In a Pocket PC project, the forms are defaulted to the landscape view and are 240 × 320; in Windows CE .NET projects, forms are 640 × 443.
- *Class library:* This creates a code library assembly that can be referenced by other projects. It's useful for encapsulating reusable code.
- *Console application (nongraphical application):* The former title appears when choosing Windows CE; the latter is for Pocket PC. In both cases, the application does contain windows, and its execution begins via a `Main` method. In the case of Windows CE, output can be directed to a console window (`Console.WriteLine`), whereas in Pocket PC the application does not support any UI.
- *Empty project:* This is an empty project that can subsequently be populated with project items.

The wizard also shows the targets that are currently available for deploying and executing the Compact Framework application. By default, SDP installs both Windows CE .NET and Pocket PC 2002 emulators and, of course, lists actual devices that may be connected to the development machine. The emulators will be discussed in more detail later in this chapter. If Pocket PC is chosen as the platform, then only the Pocket PC emulator and device will be shown.

As with other VS .NET project types, the resulting project is placed in a solution and, by default, in the current user's Visual Studio Projects directory. Although the VS .NET solution uses the same format and .sln extension, the project file extensions differ. For example, projects using VB use the .vbdproj extension while C# projects use .csdproj. It is also important to note that the solution may contain only other projects that also target the same platform for each compiler. This means that the solution cannot contain a VB project for smart devices and a VB project for the desktop Framework or even a VB project for the Pocket PC and one for Windows CE .NET. However, a VB SDP can coexist with a C# desktop Framework project. This is the case because both the Compact Framework and the desktop Framework use the same compilers with different settings, and VS .NET can instantiate only one version.

.NET Language Support

As previously mentioned, SDP supports both the VB and C# compilers and, in fact, uses exactly the same compiler as the desktop Framework, although it is targeted to the different platforms. The specifics regarding what is supported by each compiler follow.

VB

Although the VB syntax supported in SDP is not a radical subset like the VB supported in eVB, when compiling a VB Compact Framework application, there are several similarities and differences to note.

First, although VB has traditionally supported its own set of file I/O functions, including `ChDir`, `ChDrive`, `Print`, `PrintLine`, `Seek`, `FileOpen`, `FilePut`, `FileGet`, and `FileClose`, among others, these functions are no longer supported in the Compact Framework. The reason is twofold: The Compact Framework already contains all the necessary I/O functionality in the `System.IO` namespace (`File`, `Directory`, `TextReader`, `TextWriter`, and so on), and omitting the older methods reduces the size of the `Microsoft.VisualBasic.dll` assembly that must be deployed with the project.

Second, the Compact Framework does not support late binding. In other words, code like the following snippet will not compile in an SDP because of the third line:

```
Dim mine As Object
mine = Activator.CreateInstance(GetType(String))
MsgBox(mine.Length)
```

In this case, the variable mine is declared as `Object` (`System.Object`) and created using the `Activator` class. Although the instantiation works, the call to the `Length` property will cause the compilation error, "The targeted version of the .NET Compact Framework does not support late binding."

NOTE: The features of the desktop Framework class libraries, such as asynchronous delegates and ActiveX controls, are, of course, not supported in VB or in C# in the Compact Framework.

Finally, because the same compiler is used (and the EE includes a Type Checker as mentioned previously), VB in the Compact Framework supports the same set of language constructs and strict type checking using the `Option Strict On` statement as it does in the desktop Framework.

OO Comes to Devices

Although VB 6.0 and eVB developers will feel fairly comfortable with the VB syntax used in SDP, it should be noted that the object-oriented nature of the Compact Framework means that developers need to have a firm grounding in object-oriented concepts in order to maximize their productivity.

For example, the class libraries in the Compact Framework make extensive use of both implementation and interface inheritance. Developers familiar with these concepts can not only understand the framework more easily, but, using the framework, can write polymorphic code that is more maintainable and flexible than the component-based development typically done in eVB. In the long run, using object-orientation in their own designs allows developers to write less code and make the code that they do write more reusable.

To take maximum advantage of OO, however, a development organization should also get up to speed on the use of design patterns. See the "Related Reading" section for two good books that can help you get started.

C#

Unlike VB, the C# syntax used in the Compact Framework is 100% language compatible between the Compact Framework and the desktop Framework and, therefore, supports the ECMA-334 specification. This means that other than class library features that are unsupported, C# code will be easy to port from the desktop Framework to the Compact Framework and vice versa.

UI Support

Because of the obvious differences between the UI capabilities of devices targeted for the Compact Framework and desktop PCs, it is important to understand how those differences affect the design and development process for applications written for the Compact Framework and SDP.

Fortunately, although totally rewritten for the Compact Framework, the `System.Windows.Forms` namespace contains the same basic programming model (albeit a subset like the rest of the Compact Framework) as that found in the desktop Framework.[25] For example, it provides a `Control` class from which all other controls and the forms themselves are derived, just as in the desktop Framework. In addition, a forms designer is included in SDP to make development of the UI as simple as in eVB and yet more powerful. Finally, the Compact Framework and SDP also support the ability to create custom controls so that UI functionality can be easily reused across forms and projects.

NOTE: One of the features not supported by SDP is visual inheritance. It is possible to create, using this feature in VS .NET with a desktop Windows Forms application, a form complete with code and controls and then to inherit a new form from it using a menu option from within VS .NET. As changes are made to the base form, they are reflected on the derived form.

Supported Controls

To get a feel for the breadth of the supported controls, consider Table 2-3, which shows the controls included in SDP that are also in the desktop Framework.

[25] The Windows Forms assembly was rewritten primarily for performance reasons.

Table 2–3 Supported Controls in the Compact Framework

Type	Controls
Input	CheckBox, ComboBox, DomainUpDown, NumericUpDown, RadioButton, TextBox
Non-visual	Timer, InputPanel, OpenFileDialog, SaveFileDialog
Display	Label, TreeView, StatusBar, ListView, ListBox, DataGrid
Images	ImageList, PictureBox
Sliders	HScrollBar, VScrollBar, TrackBar, ProgressBar
Place holders	Panel, TabControl
Menus and Toolbars	MainMenu, ContextMenu, Toolbar

Keep in mind as well that for each of the controls listed in Table 2-3, the look and feel of the control has been designed to preserve the look and feel of the device and that a subset of the properties, methods, and events is available. For example, the **TextBox** control supports just fewer than 50% of the members of the desktop Framework.[26]

While SDP supports the majority of controls that are frequently used, there are also several controls in the desktop Framework that are not included in SDP because of device constraints, the UI design guidelines for smart devices, or a lack of support in Windows CE. These include the **GroupBox**, **RichTextBox**, **PrintDialog**, **PrintPreview**, **PrintPreview-Control**, **CheckedListBox**, **ColorDialog**, **FontDialog**, **ErrorProvider**, **HelpProvider**, **LinkLabel**, **NotifyIcon**, **Tooltip**, and **Splitter**.

The process for using controls in an SDP is the same as that in the desktop Framework. The controls can be dragged and dropped on the design surface, which creates declarations in the code-behind file for the form. Unlike desktop projects, however, the controls cannot be anchored in place on a form because forms are not typically resized on a device (especially a Pocket PC) as they are on the desktop. Developers can then manipulate the controls in the code utilizing the IntelliSense feature of VS .NET. Events are handled by double-clicking on the control, which creates an event handler in the code-behind file using the **Handles** clause in VB or creating a delegate in C#. For example, to respond to the **Click** event of a button called **btnShowStats**, the following event handler would be created in VB:

```
Private Sub btnShowStats_Click(ByVal sender As System.Object, _
   ByVal e As System.EventArgs) Handles btnShowStats.Click
```

[26] For example, none of the drag-and-drop or validation events and methods is supported.

```
    ' Perform the logic here
End Sub
```

Note that the event handler uses the same event pattern as that in the desktop Framework, where the object that raised the event and event arguments are passed into the event.

The Windows Forms designer in SDP also supports nongraphical controls through the inclusion of a panel in the designer, as shown in Figure 2-6. *Nongraphical controls are placed in a panel beneath the form, in this case a* `MainMenu` *control called* `mnuMain`.

Custom Controls

Not only can developers use the controls provided by SDP in the toolbox, they can also create their own controls by deriving them from existing controls,

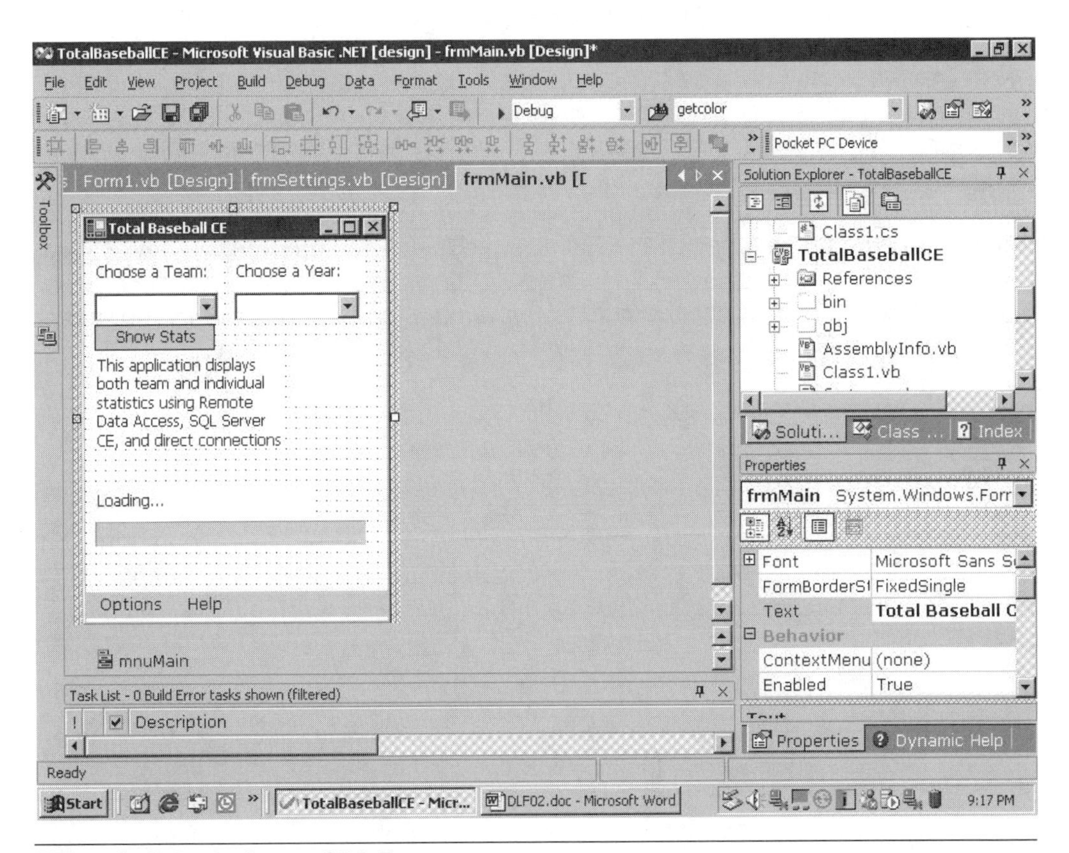

Figure 2–6 *Designing an SDP Form.*

One UI or Two?

As Figure 2-5 shows, SDP supports targeting both the Pocket PC and Windows CE platforms, and those platforms support various screen sizes with the Pocket PC at 240 × 320 and Windows CE defaulted to 640 × 443 (although the sizes will vary with the device). This difference will cause developers some difficulty when building solutions targeted at both platforms.

Although it is possible to detect the platform at runtime using an operating system function and then to rearrange and resize controls programmatically, for most applications this won't be feasible. This is primarily because moving from a larger screen size to a smaller one also entails using different graphics and even splitting controls across multiple forms, which can result in code that is difficult to maintain and extend.

As a result, a better technique is to partition the application fully using a layered architecture where the presentation code is separated from the business logic and data access. Then, create two entirely separate assemblies for the presentation code, one that targets each platform. A layer of code in each assembly (often referred to as user process components*) then mediates the calls between the business and data-access logic and the presentation code by presenting a consistent interface or API to the rest of the application. The user process objects can then fully abstract the UI differences, for example, in the name and numbers of forms.

The application can then be deployed as two separate packages (.cab files), one for each targeted device to minimize the install footprint on the device, or together in a single package. If the latter option is chosen, code in the application will need to check for the platform and to instantiate the appropriate user process components.

* Documented as the Mediator design pattern in the *Design Patterns* book referenced at the end of the chapter.

by using a combination of existing controls (referred to as composite controls), or by creating brand new controls from scratch. An example of the former, a `TextBox` that allows only numeric entry written in C#, can be seen in Listing 2-1.

Listing 2–1 *Extending an Existing Control.* This listing shows how to create a specialized version of the `TextBox` control that accepts only numeric input.

```
public class NumericTextBox : TextBox
{
```

```
    // Restricts the entry of characters to digits, the
    // negative sign, the decimal point, and editing keystrokes

    private NumberFormatInfo numberFormatInfo;
    private string groupSep;
    private string decSep;
    private string negativeSign;

    public NumericTextBox()
    {
      numberFormatInfo =
       System.Globalization.CultureInfo.CurrentCulture.NumberFormat;
      groupSep = numberFormatInfo.NumberGroupSeparator;
      decSep = numberFormatInfo.NumberDecimalSeparator;
      negativeSign = numberFormatInfo.NegativeSign;
    }

    protected override void OnKeyPress(KeyPressEventArgs e)
    {
      base.OnKeyPress(e);
      string keyInput = e.KeyChar.ToString();
      if (Char.IsDigit(e.KeyChar))
      {
        // Digits are OK
      }
      else if (keyInput.Equals(decSep) ||
        keyInput.Equals(groupSep) ||
          keyInput.Equals(negativeSign))
      {
          // Negative, decimal and group separator is OK
      }
      else if (e.KeyChar == '\b')
      {
        // Backspace key is OK
      }
      else
      {
        // Supress this invalid key
        e.Handled = true;
      }
    }
  }
```

As you can see in Listing 2-1, to extend the functionality of an existing control, a developer must simply derive from the existing control and then override the appropriate methods or add additional members when appropriate. In this case the `OnKeyPress` method is overridden to accept only digits, decimal separators, the group separator, a negative sign, and the backspace. Note also how the control in its constructor reads the current culture information to ensure that the proper decimal, separator, and negative keystrokes are accepted.

In order to create either a composite or a completely custom control where the developer determines how the control paints, the developer must derive from the `Control` class. In the latter case the developer must also override the `OnPaint` method in order to draw the control using classes from the `System.Drawing` namespace. For more information and for sample code showing a custom control see the walkthrough "Authoring a Custom Control for Smart Device Applications" in the online documentation.

KEY POINT

Whether creating a derived control as in Listing 2-1 or a custom or composite control, SDP does not currently support placing controls developed with the Compact Framework directly into the toolbox in VS .NET. In order to do so, developers must compile an alternate version of the control using the desktop Framework to be used at design time.[27] This design-time control can then be placed into the toolbox in VS .NET by right-clicking on the toolbox and selecting Add/Remove Items, as with other controls. The only caveat is that, of course, the Compact Framework must use the runtime version of the control for deployment to the device. This is accomplished by adding the `RuntimeAssemblyAttribute` to the control project and using it to specify the name, version, culture, and public key of the runtime control:

```
#if NETCFDESIGNTIME
  [assembly:
  System.CF.Design.RuntimeAssemblyAttribute("NumericTextBox,
  Version=1.0.0.0, Culture=neutral, PublicKeyToken=null")]
#endif
```

Note that the attribute is wrapped in a conditional directive that tests for the `NETCFDESIGNTIME` symbol. This symbol is specified when building

[27] An easy way to do this is to compile the control at the command line and specify the `NETCFDESIGNTIME` symbol using the `/define` command-line switch.

the design-time control and is used by the SDP forms designer to link to the runtime version, whose assembly must be placed in the CompactFrameworkSDK\v1.0.5000\Windows CE directory.

In addition, developers can use the attributes of the `System.ComponentModel` namespace of the desktop Framework to affect the appearance of properties in the custom control. For example, if a custom control contains a property called `MaxLength`, the following attributes can be placed on the property declaration to affect the property grid.

```
#if NETCFDESIGNTIME
  [System.ComponentModel.Category("Appearance")]
  [System.ComponentModel.DefaultValueAttribute(20)]
  [System.ComponentModel.Description("The maximum length.")]
#endif
```

Here the property is assigned to the Appearance category, and it is given a default value of 20 and a short description shown in the Properties window.

Components

Just as in the desktop Framework, the Compact Framework also supports creating inherently nongraphical components that can be designed graphically in the forms designer and placed in the toolbox. These components show up in the panel at the bottom of the forms designer as shown in Figure 2-6. This is accomplished by deriving a class from `System.ComponentModel.Component`. Just as with custom controls, however, a design-time version of the component assembly must be compiled. Developers must also apply the `ToolBoxItemFilterAttribute` to the class to associate the component with the SDP forms designer and to specify that the component can be used for the smart device platform.

```
<ToolBoxItemFilterAttribute("NETCF", ToolBoxItemFilterType.Require), _
  ToolBoxItemFilterAttribute("System.CF.Windows.Forms", _
  ToolBoxItemFilterType.Custom)> _
```

Just as with custom controls, the properties of the component can be marked with attributes from the `System.ComponentModel` namespace to affect the appearance of the component in the property grid.

Drawing

As mentioned previously, the Compact Framework does support the `System.Drawing` namespace to allow developers to produce graphics directly on the screen. This can be especially useful when creating custom controls. Although `System.Drawing` does not support GDI+,[28] it does support the following core drawing primitives developers expect:

- Drawing ellipses, icons, lines, images, strings, polygons, and rectangles by exposing a subset of the draw methods of the `Graphics` class
- Filling ellipses, polygons, rectangles, and regions using the fill methods of the `Graphics` class
- Providing image transparency through the inclusion of the `ImageAttribute`

As a result, although it is impossible to draw pie charts easily with the Compact Framework, it is fairly simple to draw rectangular shapes, for example, to create a bar chart, as shown in Listing 2-2, using VB.

Listing 2–2 *Drawing a Bar Chart.* This method draws a bar chart, given the PictureBox on which to draw it, the data for the X and Y axes, and the title.

```
Private Sub DrawBarChart(ByVal pb As PictureBox, _
  ByVal xAxis() As String, _
  ByVal yAxis() As Integer, ByVal title As String)

  Dim i As Integer
  Dim bm As New Bitmap(pb.Width, pb.Height)
  Dim g As Graphics
  Dim maxHeight As Integer = 180

  g = Graphics.FromImage(bm)

  ' Form color
  g.Clear(Color.Snow)

  ' Graph title
  g.DrawString(title, New Font("Tahoma", 8, FontStyle.Bold), _
    New SolidBrush(Color.Black), 5, 5)
```

[28] GDI+ is the graphics subsystem shipped with Windows XP that supports two-dimensional vector graphics, imaging, and typography and improves upon the graphics device interface (GDI) shipped with earlier versions of Windows.

```
    ' Graph legends
    Dim symbolLeg As Point = New Point(150, 10)
    Dim descLeg As Point = New Point(175, 6)
    For i = 0 To xAxis.Length - 1
      g.FillRectangle(New SolidBrush(GetColor(i)), symbolLeg.X, _
       symbolLeg.Y, 20, 10)
      g.DrawRectangle(New Pen(Color.Black), symbolLeg.X, _
       symbolLeg.Y, 20, 10)
      g.DrawString(xAxis(i).ToString, New Font("Tahoma", 8, _
       FontStyle.Regular), New SolidBrush(Color.Black), _
       descLeg.X, descLeg.Y)
      symbolLeg.Y += 15
      descLeg.Y += 15
    Next i

    ' Bars
    Dim padding As Integer = 15

    ' Find the tallest bar
    Dim max As Integer
    For Each i In yAxis
      If i > max Then
         max = i
      End If
    Next

    ' Scale the bars
    Dim yScale As Double = maxHeight / max

    For i = 0 To yAxis.Length - 1
      g.FillRectangle(New SolidBrush(GetColor(i)), _
       (i * padding) + 10, 200 - (yAxis(i) * yScale), 10, _
       (yAxis(i) * yScale) + 5)
      g.DrawRectangle(New Pen(Color.Black), (i * padding) + 10, _
        200 - (yAxis(i) * yScale), 10, (yAxis(i) * yScale) + 5)
    Next

    ' Border
    Dim p As New Pen(Color.Black)
    g.DrawRectangle(p, 1, 1, 220, 205)

    ' Set the picturebox
    pb.Image = bm

  End Sub
```

As you can see from Listing 2-2, the Compact Framework supports the `Bitmap` and `Graphics` classes necessary to draw a chart using the `Fill-Rectangle`, `DrawRectangle`, and `DrawString` methods. In this case, the method creates the graph on a `Bitmap`, and then places the bitmap on the `PictureBox` control passed into the method. To call the method, a client would then simply need to populate the required arrays and pass them to the method as follows:

```
Dim yaxis() As Integer = {755, 714, 660, 611, 499}
Dim xaxis() As String = {"Aaron", "Ruth", "Mays", "Bonds", "Sosa"}
DrawBarChart(PictureBox1, xaxis, yaxis, "Home Runs")
```

The resulting bar chart can be seen in Figure 2-7.

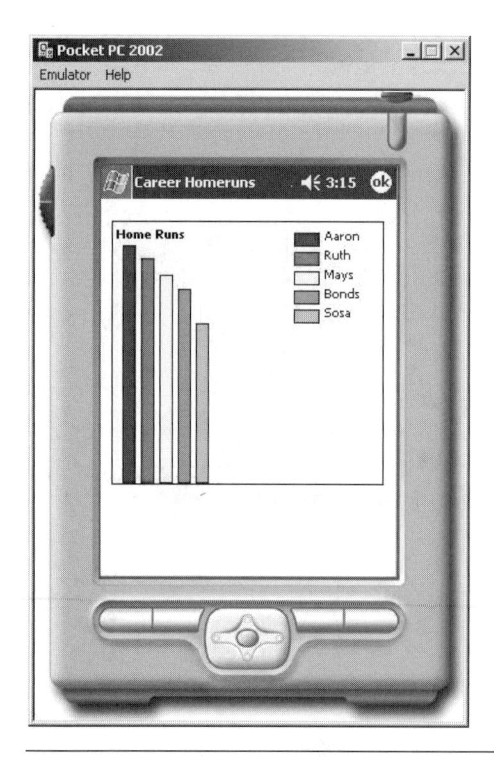

Figure 2–7 *Displaying a Bar Chart Produced in Listing 2-2 Running on the Pocket PC 2002 Emulator.*

Design Guidelines

When building solutions for the Compact Framework, it is also extremely important to take into consideration the differences between PCs and smart devices. The following are some of the important points to consider when designing the UI for applications using the Compact Framework:

- *Be consistent:* As much as is possible, follow the accepted Windows design guidelines published in documents such as the "Official Guide for User Interface Developers and Designers" published on the MSDN Web site.[29] This can ensure that your users leverage their existing knowledge. In addition, be consistent with other applications for smart devices and adopt commonly used conventions.
- *Design with user input in mind:* Because smart devices typically accept stylus input, ensure that your applications use controls that do not require character input, such as the `CheckBox`, `RadioButton`, `ComboBox`, and `NumericUpDown`. Also, be aware that on the Pocket PC, the SIP is used and covers the bottom 25% of the screen when shown. Therefore, controls that require SIP input should be repositioned so they are not covered. Finally, make controls big enough to tap easily with the stylus (a minimum of 21 × 21 pixels), and leave enough space between controls so users don't accidentally tap them.
- *Take advantage of device characteristics:* Where you can, take advantage of device-specific features, including the SIP, IrDA, and the notification API available in Pocket PC 2002.[30]
- *Promote readability:* Although designers are often tempted to use small fonts to maximize the constrained real estate available on a smart device, you should use fonts that are easily readable to prevent eyestrain.
- *Short sessions are normative:* Remember that users often use smart device in short stretches as time allows. Therefore, group required input on the initial form and allow users to come back later to add details by using menus.
- *Choose the device profile carefully:* Remember that applications created with the Windows CE profile will run on Pocket PC but not vice

---

[29] See http://msdn.microsoft.com/library/default.asp?url=/library/en-us/dnwue/html/ch00a.asp?frame=true.

[30] For an example of using the notification API, see Chapter 11.

versa. When targeting multiple profiles, add code to check for the device type, and load the appropriate forms accordingly.

- *Save defensively:* Applications running on devices such as the Pocket PC may be moved to the background simply by tapping the "X" in the corner of the window, or they may be ended by the device at any time in order to conserve resources. Therefore, your applications should save data frequently.
- *Think about navigation:* Although tempting, don't require users to do a lot of scrolling using scroll bars. Provide tabs and other more direct ways to navigate the application.

By following these guidelines, developers can create applications that perform the required functionality and that users will want to use.

Emulators

As mentioned at the beginning of this chapter, one of the goals of the Compact Framework and SDP was to provide true emulation. This goal was realized through the inclusion of two emulators, one for Windows CE .NET 4.1 and the other for Pocket PC 2002. The most interesting aspects of the emulators, however, are that they host exactly the same version of the operating system, EE, and class libraries as does managed code running on the device. This provides for a true emulation environment that allows developers to predict accurately how their applications will execute on the device.

It is important to note that the emulators are designed to run in complete isolation within the host operating system. As a result (and as you would expect), Compact Framework code running within the emulator does not have direct access to the machine on which the emulator is hosted. For example, when calling an XML Web Service hosted on the same machine as a Compact Framework application executing in the emulator, it must be called using the actual machine or Domain Name System (DNS) name, rather than localhost. In addition, the emulators have several unsurprising limitations in the following areas:

- *Networking:* Emulate only the DEC 21040 Ethernet driver and do not support the hardware and drivers for USB devices.
- *Peripherals:* Do not support any PC Card devices, Compact Flash (CF) cards, or other storage devices including CD and DVD file system drivers.
- *Display:* Do not support screen rotation or the use of multiple screens.

NOTE: As a general rule, Compact Framework code executing within the emulators will run roughly 80% as fast as the same code executing on the device.

After an SDP is created, it can be tested using an emulator simply by compiling and deploying the application in VS .NET using the Build menu. Doing so will prompt the developer to choose one of the installed devices (and to set a default device) on which to deploy the application.[31] The emulators will be installed by VS .NET as devices and, so, will always be present. In addition, the dialog may also include an actual device if the machine has been configured to synchronize with a device such as a Compaq iPaq. By choosing the emulator, it will be launched, and the Compact Framework installed (if it is not present already), followed by the application. The developer can then manually execute the application by navigating to the installation directory on the device. At this point the application will run just as if it were running on the device connected to the developer's workstation. When the developer closes the emulator, he or she will be prompted to save its state. Doing so saves the developer from redeploying the Compact Framework to the emulator the next time the application is deployed.

NOTE: In order for the SDP to be deployed in the emulator, the developer's workstation must have a valid network connection. If the workstation is disconnected from the network (as in a laptop or notebook situation), the developer will need to install the Microsoft Loopback Adapter on his workstation. This can be done by using the Add Hardware Wizard in the Control Panel and selecting Add a New Hardware Device. Add a network adapter, and choose the Microsoft Loopback Adapter from the list of adapters when Microsoft is chosen as the manufacturer.

The emulators themselves can be configured using the Options dialog found on the Tools menu within VS .NET. The Device Tools option contains both General and Devices property pages, where the developer can elect to disable the prompt to choose the deployment device and configure the emulators respectively, as shown in Figure 2-8.

The interesting aspect of the Devices property page is that it can be used to change the display, memory, and hardware settings for the default

[31] The default deployment device can also be set in the Device properties page in the Project Properties dialog.

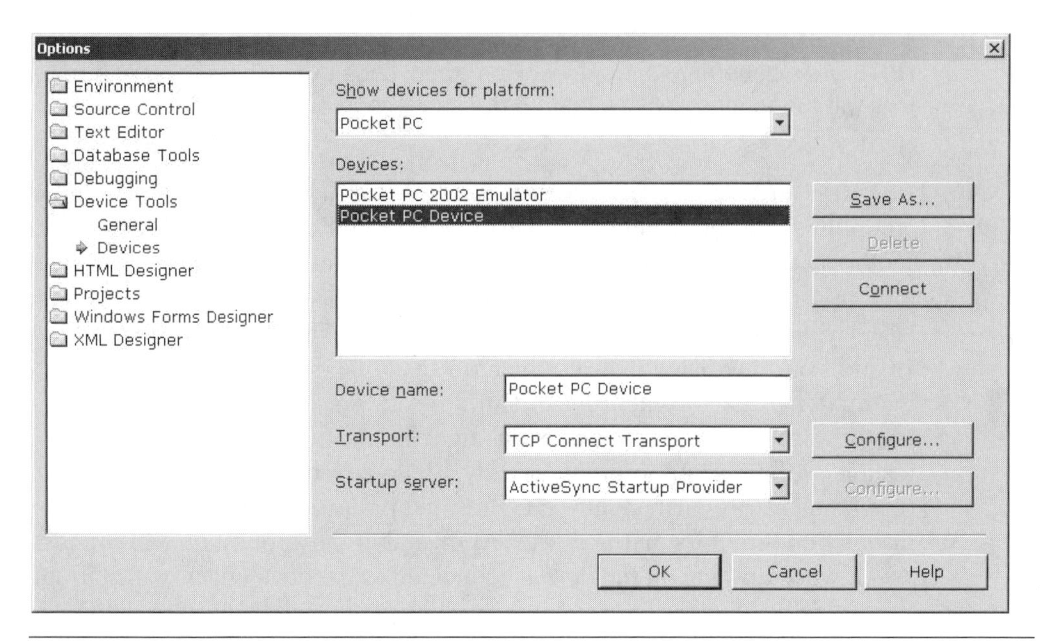

Figure 2–8 *Configuring Devices Properties*. This dialog is invoked from the Tools/Options menu in VS .NET and is used to configure the various devices available to the developer.

emulator and used to add additional emulator devices with different settings. For example, after clicking the Configure button next to the Startup Server dropdown, the dialog box shown in Figure 2-9 is displayed. This dialog allows a developer to constrain the amount of memory available to the emulator in order to test an application under different configurations.

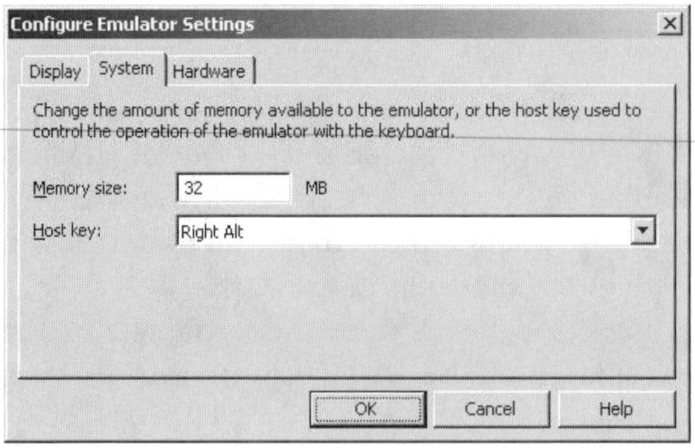

Figure 2–9 *Configuring the Emulator.*

In addition, the developer can map serial and parallel ports in the emulator to ports on the host machine, as well as change the display size and color depth. The developer can then click the Save As button to add a new device to the list of available deployment devices shown in the dialog.

Being able to change the emulator settings and create new devices in this way allows developers to test their applications effectively under a variety of conditions that emulate devices with varying displays and memory.

Outside of VS .NET, the emulators themselves can be manipulated and configured in several ways. For example, the emulators provide both hot-key and menu options for performing a pause, hard reset, soft reset, and shutdown. In addition, the hot-key combinations, based on a host key configured in the Devices property page shown in Figure 2-7, allow for displaying the emulator, help, and shortcut menus.

The emulators can also be configured with different skins analogous to the configuration of Windows Media Player. This is accomplished by creating various .bmp files and an XML document that includes the skin schema.[32]

Finally, the emulators can be executed from the command line using the `Emulator.exe` executable located in the \Program Files\Visual Studio .NET\CompactFrameworkSDK\ConnectionManager\Bin directory. Using the command line, the developer can configure the video, skin, kernel image, and Ethernet support. As mentioned previously, the emulator can include Ethernet support by emulating a single DEC 21040 Ethernet card using IP. The only requirement is that the machine hosting the emulator has an Ethernet card through which the emulator can communicate. In the event that the host machine includes multiple Ethernet cards, the Media Access Control (MAC) address of the card to be used can be specified at the command line.

To start the emulator with custom options, the following command line can be used:

```
Emulator.exe /CEImage PPC2002.bin /Video 240x320x16 /Ethernet true
```

In this case, the emulator uses the default Pocket PC kernel image, a 240×320 pixel display with Ethernet networking turned on.

---

32 To configure new emulator skins, see the emulator help file accessible from the Help menu within the emulator.

Debugging

One of the chief ways that SDP leverages VS .NET is in the use of the debugger and its tools. Because SDP can use existing tools, debugging a Compact Framework application is almost exactly like debugging a desktop Framework application and supports breakpoints, single-stepping through code, managed stack dumps that display the MSIL code, watch windows for variables, expression evaluation, and cross-language debugging (for example, stepping from an assembly written in C# into one written in VB). All of these features use the same key combinations and windows in VS .NET.

However, there are several differences:

- There is no support for viewing native instructions, registers, and call stack. As a result, if an application contains both managed and unmanaged code, the debugger steps over any unmanaged calls.
- There is no support for changing source code while the application is running (as is true of the desktop Framework).
- There is no support for attaching to a running process or Application Domain.
- Closing the device when the debugger is active causes the debugger to close with a connection failure.

Not only does SDP support debugging an application running in the emulator, it also is able to debug an application while it is executing on the device. By simply setting breakpoints and starting the debugger with the Start option on the Debug menu in VS .NET, the application is deployed to the target device, and the debugger is started. The Transport and Startup Server used to connect to the device can be configured in the Options dialog as shown in Figure 2-8 and set, for example, to use the TCP transport and the ActiveSync start-up server. For performance reasons, it is desirable to connect the device to the development machine using an Ethernet connection rather than USB.

Readers familiar with a Pocket PC will be aware that when a form appears, the icon in the upper right-hand corner may be either an X or the OK symbol. Tapping the X does not close the form, but merely sends it to the background (where it can later be closed from the Memory tab in the Settings application), although tapping OK will close the form. Developers sometimes find this behavior irritating during development because they end up opening and closing the application many times. To avoid this, a developer can place the following code in the constructor of the form:

```
#If DEBUG Then
   Me.MinimizeBox = False
#Else
   Me.MinimizeBox = True
#End If
```

This code ensures that in the Debug build, the form will display the OK button, and in Release build, the default X button will display.

As mentioned previously, to ensure that error messages are available on the device during debugging (as well as executing outside the debugger), it is necessary to reference the `System.SR.dll` assembly in the project. This ensures that the appropriate .cab file, `System_SR_language.cab`, is deployed to the device.

Finally, the command-line runtime debugger (`Cordbg.exe`), which developers can use to debug a managed application outside of VS .NET, has been augmented to support the Compact Framework through the inclusion of mode and connect arguments that allow the debugger to target device projects and connect to remote devices using a machine name and port, respectively.

Additional Tools

Although SDP contains an impressive array of development tools, there are several additional tools that often come in handy when developing SDP. Several of these tools, including a remote registry viewer and remote file system viewer for use with the emulators, are available in the eMbedded Visual Tools SDK. Many developers will want to download and install these additional tools as well.

What's Ahead

This chapter ends Part I of this book. We've attempted to put the Compact Framework in context and provide an overview of the architecture and core features that make up both the Compact Framework and SDP.

Now we're ready to move on and address the first of the essential architectural concepts needed when building business solutions: local data handling.

Related Reading

Box, Don. *Essential .NET Volume I: The Common Language Runtime*. Addison-Wesley, 2002. ISBN 0-2017-3411-7.

Fox, Dan. *Building Distributed Applications with Visual Basic .NET.* Sams, 2001. ISBN 0-672-32130-0.

Gamma, Erich, et al. *Design Patterns*. Addison-Wesley, 1995. ISBN 0-201-63361-2.

Richter, Jeffery. *Applied Microsoft .NET Framework Programming.* Microsoft Press, 2002. ISBN 0-7356-1422-9.

Shalloway, Alan, and James R. Trott. *Design Patterns Explained*. Addison-Wesley, 2002. ISBN 0-201-71594-5.

Wigley, Andy, and Stephen Wheelwright. *Microsoft .NET Compact Framework, Core Reference*. Microsoft Press, 2003. ISBN 0-7356-1725-2.

The Microsoft .NET initiative home page, at www.microsoft.com/net.

The .NET Compact Framework team page on GotDotNet, at www.gotdotnet.com/team/netcf.

The Official Microsoft Smart Devices Developer Community, at http://smartdevices.microsoftdev.com.

See Quilogy's Atomic Group for technical information and news on .NET and emerging technologies, at http://atomic.quilogy.com.

ECMA, at www.ecma.ch.

Microsoft's Shared Source Implementation of the CLI, referred to as "Rotor," at www.123aspx.com/rotor/default.aspx.

Fox, Dan. "Solve Common Design Problems," *.NET Magazine* (October 2002), at www.fawcette.com/dotnetmag/2002_10/magazine/features/dfox.

PART II

Essential Architectural Concepts

CHAPTER 3

Accessing Local Data

Executive Summary

It is axiomatic that applications running on smart devices need to have robust capabilities for local data handling. This is the case because at least some of the time (and for some applications, all of the time), devices will be disconnected from any external network or data source. That is why local data handling is the first essential architectural concept addressed in this book.

As a result, it is imperative that any programming model for smart devices include three fundamental services, including programmatic access, or the ability to work with data via code; persistence, or the ability to save and retrieve data directly on the device; and UI support, or the ability to display, edit, and validate data in a straightforward fashion. Fortunately, for organizations looking to build solutions on the Compact Framework and SDP, there is plenty of functionality built into the product to support files, XML, and relational data, not to mention custom objects.

For example, to access files and folders on a device, developers can use the classes in the `System.IO` namespace that include stream readers and writers used to read and write plaintext and binary files, as well as classes to manipulate files and folders directly. These can be augmented with some simple calls to the Windows CE API in order to retrieve system file paths as well. The Compact Framework, although not supporting all the asynchronous capabilities of its desktop cousin, even supports asynchronous I/O through direct thread manipulation and an elegant technique for updating the UI after work has been completed.

To handle XML, the classes in the `System.Xml` namespace can be used in both a streamed fashion through a reader and writer, as well as through the familiar Document Object Model (DOM), although the cursor-based reader and writer are faster and consume less memory on the device.

Relational data can be manipulated locally (perhaps after being retrieved from a remote source) using the `DataSet`, `DataTable`, and associated objects in the `System.Data` namespace commonly referred to as

ADO.NET. The methods of the data set particularly can be used to read and write data and track changes between application sessions on the device using an XML grammar referred to as a DiffGram.

Finally, UI support is included in the Compact Framework through both simple and complex data binding. Data from objects including the `DataTable` and `DataView`, as well as arrays and custom collection objects supporting the `IList` interface, can be bound to the `DataGrid`, `ListBox`, and `ComboBox` controls, as well as simpler controls, such as the `TextBox`. When the control does not support binding, a manual binding approach may also be used.

The end result is that the Compact Framework adroitly provides all the essential support for handling data locally on the device.

The Need for Local Data Handling

The first essential architectural concept, and the one addressed in this chapter, is the need to manipulate data directly on the device. This concept is fundamental because although smart device applications often retrieve and update data using a variety of remote mechanisms, including XML Web Services, HTTP, Microsoft SQL Server, IrDA, and direct socket-based communication, as discussed in the next chapter, those applications also typically require three key services:

1. *Programmatic access:* Smart device applications need to be able to work programmatically with data in several ways using a simple and consistent set of APIs. Fortunately, the Compact Framework includes subsets of the `System.Xml`, `System.Data`, and `System.IO` namespaces that allow developers to leverage their knowledge of the desktop Framework and access a robust set of functionalities for manipulating XML, relational, and file-based data directly on the device.
2. *Persistence:* Because many smart device applications will be used in a disconnected or occasionally connected mode, as discussed in Chapter 1, it is important to be able to persist the data locally on the device so that it can be reloaded if the application is closed. Once again, the namespaces of the Compact Framework provide simple mechanisms to store and retrieve XML, relational, and file-based data on the device.[1]

[1] More sophisticated persistance using SQLCE 2.0 will be discussed in Chapter 5. This chapter will not discuss synchronizing data with back-end data stores, which will be handled in Chapters 6 and 7.

3. *UI support:* Once the data is retrieved, it is usually displayed to the user in a form for viewing or editing. Fortunately, SDP includes a fairly complete set of controls, as discussed in Chapter 2. Some of these controls can even be used to bind data directly, thereby decreasing the need to write code.

The remainder of this chapter will discuss the three parts of the Compact Framework that allow developers to work with file-based, XML, and relational data locally on the device in the context of the first two services discussed earlier (programmatic access and persistence). The final section will detail how data can be displayed on the devices.

Using File I/O

Desktop Framework developers will be aware that programmatic access to file I/O is found in the `System.IO` namespace. This is also true of the Compact Framework, where I/O is encapsulated by abstracting the concept of a stream used to read and write data from the "backing store," or a medium used to store the data. Because of this abstraction, you can think of the `System.IO` namespace as consisting of three logical components, as shown in Figure 3-1.

For eVB developers, this programming model may take a little getting used to because the Compact Framework does not support the `FileSystem` class in the `Microsoft.VisualBasic` namespace that includes analogs to many of the statements and functions historically used by VB developers, such as `FileOpen`, `Input`, `LineInput`, and so on.

Foundational not only to file access but to dealing with all types of I/O is the `Stream` class found in the `System.IO` namespace. `Stream` is a base class that represents the stream of bytes to be read from or written to a backing store. As a result, it includes methods to perform these operations both synchronously (`Read`, `Write`) and asynchronously (`BeginRead`, `BeginWrite`, `EndRead`, `EndWrite`). However, while these methods are present, if developers attempt to use the asynchronous methods in the Compact Framework, a `NotSupportedException` will be thrown. The `Stream` also exposes methods that manipulate the current position in the `Stream`, such as `Seek`, and a variety of properties to interrogate the capabilities of the `Stream`, such as `CanRead`, `CanSeek`, and `CanWrite`. Because `Stream` is a base class, the `System.IO` namespace includes two classes that inherit from it to support specific backing stores. The `FileStream` class supports stream operations against physical files, whereas the `MemoryStream` class supports

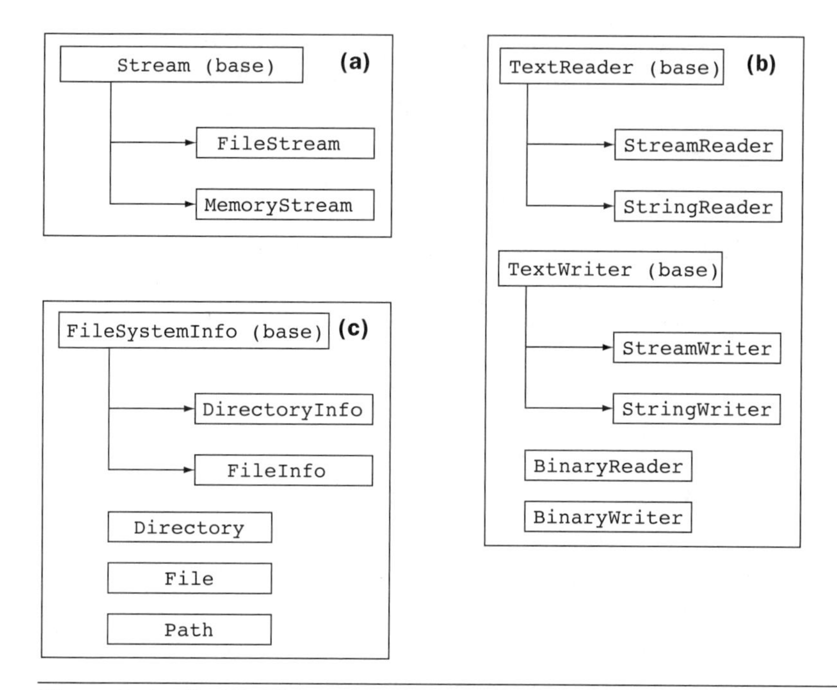

Figure 3–1 *File I/O.* The major classes of the `System.IO` namespace are broken down into components, including (a) streams, (b) readers and writers, and (c) file system classes.

stream access to physical memory. In addition, the `System.Net.Sockets` namespace implements the `NetworkStream` class to provide the underlying stream of data for network access.[2] Obviously, by deriving these classes from `Stream`, developers can take advantage of polymorphism to write more reusable and maintainable code.

The second component of `System.IO` includes the `Reader` and `Writer` classes. As the names imply, these classes are used to read and write bytes to and from a `Stream` in a particular way. Although developers can use the `Read` and `Write` methods of the `Stream` classes directly, doing so means having to read and write data as byte arrays using offsets.

There are two basic divisions in the `Reader`/`Writer` classes that include the `TextReader` and `TextWriter` classes and the `BinaryReader` and `BinaryWriter` classes. `TextReader` and `TextWriter` are base classes that read and write individual text characters to a stream, while their analogs read and write primitive types in binary. In turn, the `TextReader` and

[2] The `BufferedStream` class is not supported in the Compact Framework.

TextWriter serve as the base classes for the StreamReader and String-Reader and the StreamWriter and StringWriter classes, respectively. The StreamReader and StreamWriter classes read and write a variety of data types (including text) to a Stream in a particular encoding, whereas the StringReader and StringWriter simply read and write from strings using a StringBuilder from the System.Text namespace.

The final component of System.IO includes the various classes that deal specifically with the file system and interact with the FileStream class. As Figure 3-1 shows, these include the DirectoryInfo and FileInfo classes derived from FileSystemInfo used to manipulate files and directories in conjunction with a FileStream. In addition, the sealed Directory, File, and Path classes aid in the creation of file system objects, in addition to providing methods to copy, delete, open, and move files and directories.[3]

Reading and Writing Text Files

To illustrate the use of the System.IO namespace to read and write individual files, consider the code in Listings 3-1 and 3-2, where the methods write and read baseball box score information to and from a comma-delimited text file, respectively. This is the kind of code that a developer would write for a stand-alone application that allowed users to score a baseball game.

Listing 3–1 *Writing to a Text File.* This listing shows how a developer would use the FileStream and StreamWriter classes to write to the file system.

```
Public Sub SaveToCSV(ByVal fileName As String)

    ' Save the current box score to a CSV file

    Dim fs As FileStream
    Dim sr As StreamWriter

    Try
        ' Overwrite file if exists
        fs = New FileStream(fileName, FileMode.Create, _
          FileAccess.Write, FileShare.None)
        ' Associate the stream writer with the file
        sr = New StreamWriter(fs, System.Text.Encoding.Default, 1024)
```

[3] The FileSystemWatcher class that accompanies these classes in the desktop Framework is not included in the Compact Framework.

```
Catch e As IOException
    Throw New Exception("I/O error. Cannot access the file " & _
      fileName & " :" & e.Message)
End Try

Try
    ' Write the header
    sr.WriteLine(fileName & " created on " & _
      DateTime.Now.ToShortDateString)
    sr.WriteLine(Me.GameDate & "," & Me.GameTime)

    ' Write visiting team
    sr.WriteLine(Me.Visitor)
    Dim p As PlayerLine
    For Each p In Me.VisitingPlayers
        sr.WriteLine(p.ToString(","))
    Next
    sr.WriteLine("END")

    ' Write home team
    sr.WriteLine(Me.Home)
    For Each p In Me.HomePlayers
        sr.WriteLine(p.ToString(","))
    Next
    sr.WriteLine("END")

    ' Write the line score
    Dim i As Integer
    For i = 1 To Me.VisitingLine.Count
        sr.Write(VisitingLine(i))
        If i < Me.VisitingLine.Count Then
            sr.Write(",")
        End If
    Next
    sr.WriteLine()
    For i = 1 To Me.HomeLine.Count
        sr.Write(HomeLine(i))
        If i < Me.HomeLine.Count Then
            sr.Write(",")
        End If
    Next

Catch e As Exception
    Throw New ApplicationException("Could not write box score", e)
Finally
```

```
        sr.Close()
    End Try

End Sub
```

Listing 3–2 *Reading from a Text File.* This listing shows how a developer would use the `FileStream` and `StreamReader` classes to read to a file on the file system.

```
Public Sub LoadFromCSV(ByVal fileName As String)
    ' Read the box score from a CSV file

    Dim fs As FileStream
    Dim sr As StreamReader

    Try
        ' Read the file
        fs = New FileStream(fileName, FileMode.Open, _
          FileAccess.Read, FileShare.Read)
        ' Associate the stream writer with the file
        sr = New StreamReader(fs, True)
    Catch e As IOException
        Throw New Exception("I/O error. Cannot access the file " & _
          fileName & " :" & e.Message)
    End Try

    Try
        ' Skip the header
        sr.ReadLine()

        Dim info As String = sr.ReadLine()
        Dim gameInfo() As String = info.Split(",")
        Me.GameDate = gameInfo(0)
        Me.GameTime = gameInfo(1)

        ' Read visiting team
        Me.Visitor = sr.ReadLine()
        Dim pstr As String = sr.ReadLine()
        Do While pstr <> "END"
            Dim p As New PlayerLine(pstr, ",")
            Me.VisitingPlayers.Add(p)
            pstr = sr.ReadLine()
        Loop
```

```
            ' Read home team
            Me.Home = sr.ReadLine()
            pstr = sr.ReadLine()
            Do While pstr <> "END"
                Dim p As New PlayerLine(pstr, ",")
                Me.HomePlayers.Add(p)
                pstr = sr.ReadLine()
            Loop

            ' Read the line score
            Dim line As String = sr.ReadLine()
            Dim lines() As String = line.Split(",")
            Dim i As Integer
            For i = 0 To lines.Length - 1
                Me.VisitingLine.Add(i + 1, lines(i))
            Next
            line = sr.ReadLine()
            lines = line.Split(",")
            For i = 0 To lines.Length - 1
                Me.HomeLine.Add(i + 1, lines(i))
            Next

        Catch e As Exception
            Throw New ApplicationException( _
              "Could not read in the box score", e)
        Finally
            sr.Close()
        End Try

    End Sub
```

In Listing 3-1 you can see that the `SaveToCSV` method accepts the filename as a parameter and uses it to overwrite an existing file of the same name by passing the `FileMode.Create` value to the constructor of the `FileStream` class. A `StreamWriter` that points to the stream to write to is then instantiated. Note that the default encoding (in this case UTF-8) is used along with a buffer size of 1K, the default being 4K. If the file cannot be accessed, an `IOException` will be thrown and the method terminates.

The remainder of the `SaveToCSV` method uses the overloaded `Write` and `WriteLine` methods of the `StreamWriter` class to write out box score data. In this case the SaveToCSV method exists in a class called Scoresheet that exposes the following public fields:

```
Public VisitingPlayers As ArrayList
Public HomePlayers As ArrayList
Public HomeLine As ListDictionary
Public VisitingLine As ListDictionary
Public Home, Visitor, GameDate, GameTime As String
```

The individual player's statistics are stored as instances of the PlayerLine class in the VisitingPlayers and HomePlayers collections. The PlayerLine class contains a ToString method that accepts a delimiter that then creates a delimited string with all of the player's statistics. The end result is a comma-delimited file that contains the entire box score for a baseball game.

NOTE: Calling the Close method of the StreamWriter class in the Finally block ensures that all data is written to the stream and, hence, to the file before it is closed.

In Listing 3-2 the reverse process occurs in the LoadFromCSV method, and the comma-delimited file is loaded into a FileStream object and read with the StreamReader class. In this case developers can rely on the Split method of the String class to create arrays from the delimited data and then parse the arrays into the correct data structure. In fact, the PlayerLine class includes an overloaded constructor that accepts the delimited string, along with the delimiter, and then parses it and loads it into its public properties.

Asynchronous File Access

KEY POINT

As mentioned previously, the Stream class and its descendants, such as FileStream, expose four methods used in the desktop Framework for asynchronous reading and writing. However, these methods throw a NotSupportedException when used in the Compact Framework.

As an alternative, developers can manipulate threads directly using the classes of the System.Threading namespace. Although discussed in more detail in the next chapter, the entire SaveToCSV method shown in Listing 3-1 could be invoked on a background thread as follows:

```
Dim t As New Thread(AddressOf MySaveToCSV)
outFile = "boxscore.txt"
t.Start()
```

Doing so allows the user to continue with other useful work while the file is being saved. Of course, because the method used as the address at which to begin the thread cannot accept arguments, the `MySaveToCSV` method actually makes the call to the `SaveToCSV` method, passing in the `outFile` variable as an argument, as shown here:

```
Public Sub MySaveToCSV()
    s.SaveToCSV(outFile)
End Sub
```

Unfortunately, the Compact Framework does not support the `IsAlive`, `IsBackground`, or `ThreadState` properties or the `Interrupt` and `Join` methods, which could all be used to determine whether the thread was still executing. However, even if all open windows in the application are closed, the thread will continue to execute and the application will not be unloaded until execution completes. If the `Exit` method of the `Application` class is called, all windows and the application itself will not be unloaded until the thread completes. This behavior is different from the desktop Framework

Synchronizing Access to Resources

While the `SaveToCSV` method is executing on the background thread, the developer would also need to ensure that other threads do not access instance data that is critical to the execution of the method. To do so, the developer can use the `Monitor` class in the `System.Threading` namespace or the `SyncLock` and `lock` statements in VB and C# respectively. However, the recommended way to execute processes on separate threads that do not require access to shared resources is as follows:

1. Encapsulate the process that is to be run in a class that exposes an entry point used to start the process and instance variables to handle the state.
2. Create a separate instance of the class.
3. Set any instance variables required by the process.
4. Invoke the entry point on a separate thread.
5. Do not reference the instance variables of the class.

where threads set with low priorities (`BelowNormal` or `Lowest`) will not finish executing if the application is shut down.

In many cases it is also desirable to notify the main window running on a foreground thread when a background thread such as that shown earlier has completed. This would be the case, for example, if the `LoadFromCSV` method were invoked on a background thread that needed to update the UI when the `Scoresheet` class was loaded. While this functionality was built into asynchronous delegates not accessible in the Compact Framework, this can easily be accomplished with delegates directly. For example, assume that the `LoadFromCSV` method is to be called on a background thread. The form that makes the call can include a form-level `EventHandler` delegate declared as follows:

```
Private UICallback As EventHandler
```

Then, before the load method is invoked on the thread, the delegate is instantiated to point to a method called `LoadUI` on the form. This method is responsible for updating controls on the UI with the score sheet data.

```
UICallback = New EventHandler(AddressOf LoadUI)
Dim t As New Thread(AddressOf MyLoadFromCSV)
inFile = "boxscore.txt"
t.Start()
```

Finally, when the loading is completed, the `MyLoadFromCSV` method can invoke the delegate that points to the `LoadUI` method. The only caveat is that the `LoadUI` method must use the standard `EventHandler` delegate that accepts an object (the sender) and an object of type `EventArgs`.

```
Public Sub MyLoadFromCSV()
    s.LoadFromCSV(inFile)
    Me.Invoke(UICallback)
End Sub
```

By invoking the delegate on `Me` (`this` in C#, meaning the current form), the Compact Framework ensures that the `LoadUI` method will be executed safely on the foreground thread. If a developer attempts to update the UI running on the foreground thread directly from code running on the background thread, the application will hang.

Unfortunately, the Compact Framework does not support the overloaded `Invoke` method, which accepts arguments. However, a developer could wrap this functionality in his or her own class, such as the `Invoker` class shown in Listing 3-3.

Listing 3–3 *Updating the UI with an* `Invoker` *Class.* This class can be used to update the UI of a form in the Compact Framework and pass it arguments.

```
Public Delegate Sub UIUpdate(ByVal args() As Object)
Public Class Invoker

    Private _control As Control
    Private _uiUpdate As UIUpdate
    Private _args() As Object

    Public Sub New(ByVal c As Control)
        ' Store the control that is to run the method on its thread
        _control = c
    End Sub

    Public Sub Invoke(ByVal UIDelegate As UIUpdate, _
         ByVal ParamArray args() As Object)
        ' called by the client and passed the delgate that
        ' points to the method to run
        ' as well as the arguments
        _args = args
        _uiUpdate = UIDelegate
        _control.Invoke(New EventHandler(AddressOf _invoke))
    End Sub

    Private Sub _invoke(ByVal sender As Object, ByVal e As EventArgs)
        ' this is now running on the same thread as the control
        ' so freely call the delegate
        _uiUpdate.Invoke(_args)
    End Sub

End Class
```

Here the client simply needs to create an instance of `Invoker` and pass it the control (such as the form) on which to execute the method.

```
Private inv As Invoker
inv = New Invoker(Me)
```

Then, when the method running on the other thread completes, it can simply call the **Invoke** method, passing in the delegate that contains the method to update the UI, along with the arguments.

```
inv.Invoke(New UIUpdate(AddressOf LoadUI), "1", "2")
```

Obviously, the asynchronous techniques discussed in this section could also apply to working with XML and relational data, covered in the following sections.

Manipulating Files and Directories

The Compact Framework also supports manipulating files and folders directly using the `File`, `FileInfo`, `DirectoryInfo`, and `Directory` classes in the `System.IO` namespace. Like their desktop Framework equivalents, these classes allow a developer to enumerate and inspect files and directories and copy, move, and delete them. For example, the method in Listing 3-4 uses these classes to move all the files matching specific criteria to an archive directory relative to the path.

Listing 3–4 *Manipulating Files and Directories.* This method uses the classes of System.IO to move files from one directory to another.

```
Public Sub ArchiveFiles(ByVal filePath As String, _
  ByVal criteria As String)

    Dim f As FileInfo
    Dim dDir As DirectoryInfo
    Dim dArchive As DirectoryInfo

    ' Make sure directory exists
    If Not Directory.Exists(filePath) Then
        Throw New ApplicationException("Directory " & filePath & _
           " does not exist")
    Else
        dDir = New DirectoryInfo(filePath)
    End If

    ' Create the archive directory
    dArchive = Directory.CreateDirectory(filePath & _
      Path.DirectorySeparatorChar & "Archive")
```

```
    ' Get all the files in the directory
    If criteria Is Nothing Then
       criteria = "*.*"
    End If

    For Each f In dDir.GetFiles(criteria)
        Try
            ' Move the file and delete
            f.MoveTo(dArchive.FullName & _
              Path.DirectorySeparatorChar & f.Name)
         Catch e As Exception
           Throw New ApplicationException("Error on file " & f.Name, e)
        End Try
    Next
End Sub
```

It is interesting to note that the `File` and `Directory` classes are used statically to manipulate objects in the file system, whereas the `DirectoryInfo` and `FileInfo` classes represent individual file system entries. In other words, the methods of `File` and `Directory` can be used to perform operations on files and directories, whereas classes derived from `FileSystemInfo` represent specific instances of files and directories. In addition, the `File` and `Directory` classes accept `String` arguments and return arrays of strings when queried for data, for example, using the `GetFiles` method shown in Listing 3-4, whereas the `FileSystemInfo` classes accept and return other instances of a `FileSystemInfo` class. The `Path` class also is used statically to return platform-independent delimiters and other information as shown through the use of the `DirectorySeparatorChar` property.[4]

KEY POINT

Although the file and directory classes implement most of the functionality of the desktop Framework, the `Directory` class's `GetCurrentDirectory` method throws a `NotSupportedException` instead of returning the working directory of the application. However, the current directory and other system folders can be retrieved using a simple wrapper class like that shown in Listing 3-5.

[4] There is also some overlap between the `File` and `FileInfo` classes and the `FileStream` discussed previously. For example, a developer can use the shared methods `OpenText`, `CreateText`, or `AppendText` of the `File` and `FileInfo` classes to open a text file as well.

Listing 3–5 *Finding System Folders.* This wrapper class allows easy access to Windows CE system folders and the folder that the application is executing from.

```
Namespace Atomic.CEUtils

 Public Enum ceFolders As Integer
    PROGRAMS = 2          ' \Windows\Start Menu\Programs
    PERSONAL = 5          ' \My Documents
    STARTUP = 7           ' \Windows\StartUp
    STARTMENU = &HB       ' \Windows\Start Menu
    FONTS = &H14          ' \Windows\Fonts
    FAVORITES = &H16      ' \Windows\Favorites
 End Enum

 Public Class FileSystem

    Private Sub New()
    End Sub

    Private Const MAX_PATH As Integer = 260
    Private Shared _specialFolderPath As String
    Private Shared _documentsFolder As String
    Private Shared _windowsFolder As String
    Private Shared _assemblyFolder As String

    <DllImport("coredll.dll")> _
     Private Shared Function SHGetSpecialFolderPath( _
            ByVal hwndOwner As Integer, _
            ByVal lpszPath As String, _
            ByVal nFolder As ceFolders, _
            ByVal fCreate As Boolean _
            ) As Boolean
     End Function

    Public Shared ReadOnly Property WindowsFolder() As String
        Get
          If _windowsFolder Is Nothing Then
              _windowsFolder = _getWindowsFolder()
          End If
          Return _windowsFolder
        End Get
    End Property
```

```
    Public Shared ReadOnly Property DocumentsFolder() As String
        Get
          If _documentsFolder Is Nothing Then
              _documentsFolder =
GetSpecialFolderPath(ceFolders.PERSONAL)
          End If
          Return _documentsFolder
        End Get
    End Property

    Public Shared ReadOnly Property RuntimeFolder() As String
        Get
          If _assemblyFolder Is Nothing Then
              Dim a As [Assembly]
              a = System.Reflection.Assembly.GetExecutingAssembly()
              _assemblyFolder = a.GetName().CodeBase
              _assemblyFolder = _assemblyFolder.Substring(0, _
                  _assemblyFolder.LastIndexOf("\"))
            End If
            Return _assemblyFolder
        End Get
    End Property

    Public Shared Function GetSpecialFolderPath( _
       ByVal folder As ceFolders) As String
        Dim sPath As String = New String(" "c, MAX_PATH)
        Dim i As Integer

        Try
            SHGetSpecialFolderPath(0, sPath, folder, False)
            i = sPath.IndexOf(Chr(0))
            If i > -1 Then
                sPath = sPath.Substring(0, i)
            End If
        Catch ex As Exception
            sPath = ex.Message
        End Try
        Return sPath
    End Function

    Private Shared Function _getWindowsFolder() As String
        Dim s As String

        s = GetSpecialFolderPath(ceFolders.STARTMENU)
        Dim i As Integer = s.LastIndexOf("\")
```

```
        If i > -1 Then
            s = s.Substring(0, i)
        End If
        Return s
    End Function


 End Class
End Namespace
```

As you can see in Listing 3-5 the `FileSystem` class in the `Atomic.CEUtils` namespace includes shared properties to return the documents and windows folders by calling the Windows CE API function `SHGetSpecialFolderPath` found in `coredll.dll`. This is an example of using the PInvoke functionality of the Compact Framework to make direct calls to the underlying operating system. This function is also exposed through the `GetSpecialFolderPath` method that accepts one of the `ceFolders` enumerated types. The runtime folder, however, is accessed through the `GetExecutingAssembly` method of the `System.Reflection` namespace.

Finally, it is also possible to use the `OpenFileDialog` class of the `System.Windows.Forms` namespace to open a dialog from which the user can select a file and return it. There are several differences, however, in its operation in the Compact Framework. For example, although it supports the `InitialDirectory` property, setting it has no effect, and the dialog will always display all of the documents in the My Documents folder.[5] The folder dropdown list can then be used to filter based on the folder within My Documents. Also, the Compact Framework does not support multifile selection, filtering on read-only files, checking for the file's existence, and opening the file when selected. As a result, typical usage of this class is as follows:

```
Dim f As New OpenFileDialog

f.Filter = "All files (*.*)|*.*|Scoresheet files (*.scr)|*.scr"
If f.ShowDialog() = DialogResult.OK Then
    _doSomeWork (f.FileName)
End If
```

[5] This is also the behavior that occurs when calling the `GetOpenFileName` Windows CE API function, even if a directory path is specified in the structure passed to the function.

Handling XML

As mentioned in the previous chapter, the Compact Framework includes a subset of the support for handling XML found in the desktop Framework within the `System.Xml` namespace. Particularly, this means that the Compact Framework ships with the `System.Xml.Schema` and `System.Xml.Serialization` namespaces (with the most significant omission being the `XmlSerializer` class), but not the `System.Xml.XPath` and `System.Xml.Xsl` namespaces. In addition, although the `System.Xml.Schemas` namespace is included with its `XmlSchemaObject`, `XmlSchema`, and `XmlSchemaException` classes, these classes are not functional and cannot be used to load, manipulate, and save XML Schema Definition Language (XSD) documents. However, even with these omissions, developers will find a wealth of functionality for reading and writing XML documents using both the DOM and stream-based readers and writers.

NOTE: Although XML Stylesheet Transformations (XSLT) are not supported, keep in mind that XSLT is particularly useful in Web programming environments, where server-side code is called upon to transform an XML document using an XSL stylesheet. Because the Compact Framework does not support ASP.NET, there is little need for XSLT.

Using the DOM

The `System.Xml` namespace includes the familiar DOM programming model through its `XmlDocument` class, which implements the W3C DOM Level 1 Core and the Core DOM Level 2 specifications using an in-memory tree representation of the document. Many developers are already familiar with the DOM from working with the COM-based Microsoft XML Parser (MSXML), and so, using `XmlDocument` to manipulate local XML will often represent the smallest learning curve. In fact, the `XmlDocument` class is analogous to the `DOMDocument` class found in MSXML.

To load an XML document with the `XmlDocument` class, a developer can use either the `LoadXml` or the `Load` method. The former accepts a string that includes the well-formed XML to load, while the latter is overloaded to accept a filename, an `XmlReader`, a `TextReader`, or a `Stream` object. If the XML from the source is not well formed, an `XmlException` will be thrown. For example, consider an XML document that represents a score sheet for a baseball game, part of which (only the first batter and first at bat for each team, due to space limitations) is shown in Listing 3-6.

Listing 3–6 *A Sample XML Document.* This XML document represents a score sheet from a baseball game. Note that the `Player` and `PA` elements would repeat in a typical document.

```
<?xml version="1.0" encoding="utf-8"?>
<Scoresheet Visitor="Chicago Cubs" Home="San Francisco Giants"
  Date="08/06/2002" Time="9:15 PM" >
  <Visitor>
    <Lineup>
      <Player Order="1" Name="Mark Bellhorn" Position="2B"
        Inning="1" />
    </Lineup>
    <PA Inning="1" Order="1" Pitches="BBS" OutNumber="1" Out="F7" />
  </Visitor>
  <Home>
      <Lineup>
          <Player Order="1" Name="Lofton" Position="CF" Inning="1" />
      </Lineup>
     <PA Inning="1" Order="1" Pitches="BBS" LastBase="1" Result="1B" /
  >
  </Home>
</Scoresheet>
```

This document can be loaded by the `Scoresheet` class discussed previously and parsed using the code in the `LoadXmlDoc` method in Listing 3-7. Note that the familiar `DocumentElement` and `GetElementsByTagName` members are present, as in other implementations of the DOM, such as MSXML. Developers then work with the individual elements in the document using the `XmlNode` class, which serves as the base class for `XmlDocument` and other classes, such as `XmlAttribute`, which presents attributes.

Listing 3–7 *Manipulating a Document with the DOM.* This method loads an XML document using the DOM. The `_addPlayers` method parses an entire `XmlNodeList` for the home and visiting teams.

```
Public Sub LoadXmlDoc(ByVal fileName As String)
    Dim d As New XmlDocument

    Try
        ' Load the xml file
        d.Load(fileName)
```

```
        Me.Visitor = d.DocumentElement.Attributes("Visitor").Value
        Me.Home = d.DocumentElement.Attributes("Home").Value
        Me.GameDate = d.DocumentElement.Attributes("Date").Value
        Me.GameTime = d.DocumentElement.Attributes("Time").Value

        Dim xnl As XmlNodeList

        ' Parse the visiting team
        xnl = d.GetElementsByTagName("Visitor")
        _addPlayers(xnl, Me.VisitingPlayers, Me.VisitingLine)
        ' Parse the home team
        xnl = d.GetElementsByTagName("Home")
        _addPlayers(xnl, Me.HomePlayers, Me.HomeLine)

    Catch e As XmlException
       Throw New ApplicationException("Could not load " & fileName, e)
    End Try
End Sub
```

In order to persist an XML document loaded into the DOM, developers can use the `Save` method of the `XmlDocument` class. This method is overloaded and allows saving to a file, a `Stream`, a `TextWriter`, or an `XmlWriter` (to be discussed shortly). In this way, developers have the flexibility to use an already existing stream or even to store the XML in memory using a `MemoryStream` for a short period.

Using XML Readers and Writers

One of the most interesting innovations supported by the desktop Framework and carried into the Compact Framework is the way developers can interact with XML documents through the use of a stream-based API analogous to the stream reading and writing performed on files. At the core of this API are the `XmlReader` and `XmlWriter` classes, which provide read-only, forward-only, cursor-style access to XML documents and a mechanism for writing out XML documents, respectively. Because these classes implement a stream-based approach, they do not require that the XML document be parsed into a tree structure and cached in memory as happens when working with the document through the `XmlDocument` class.

Using XML Readers

Obviously, the DOM programming model is not ideal for all applications, particularly when the XML document is large. Any but the smallest XML documents have both the effect of slowing performance because of having to build the DOM tree and consuming additional memory to store the tree. On CPU and memory-constrained devices like those running the Compact Framework, this is especially important to consider.[6]

The `XmlReader` is designed to alleviate these constraints by combining the best aspects of the DOM and the event-based Simple API for XML (SAX) API in MSXML in the context of a stream-based architecture. In this model developers pull data from the document using an intuitive cursor-style looping construct, rather than simply being pushed data by responding to events fired from the parser or querying an already existing tree structure.

The `XmlReader` class is actually an abstract base class for the `XmlTextReader`, and `XmlNodeReader` classes and is often used polymorphically as the input or output arguments for other methods in the Compact Framework. An example of using the `XmlTextReader` to parse the XML document shown in Listing 3-6 is produced in Listing 3-8. Notice that this listing is functionally identical to Listing 3-7.

Listing 3–8 *Manipulating a Document with an `XmlReader`.* This method loads the same XML document as in Listing 3-6, but this time using an `XmlTextReader`.

```
Public Sub LoadXmlReader(ByVal fileName As String)

    Dim xlr As XmlTextReader

    Try
       xlr = New XmlTextReader(fileName)
       xlr.WhitespaceHandling = WhitespaceHandling.None

       Do While xlr.Read()
           Select Case xlr.Name
               Case "Scoresheet"
                   If xlr.IsStartElement Then
```

[6] To address these issues on desktop PCs, Microsoft included Simple API for XML (SAX) in MSXML 3.0 to provide an event-driven programming model for XML documents. Although this alleviated the performance and memory constraints of the DOM, it did so at the cost of complexity.

```
                    Me.Home = xlr.GetAttribute("Home")
                    Me.Visitor = xlr.GetAttribute("Visitor")
                    Me.GameTime = xlr.GetAttribute("GameTime")
                    Me.GameDate = xlr.GetAttribute("GameDate")
                End If
            Case "Visitor"
                If xlr.IsStartElement Then
                    _addPlayersReader(xlr, Me.VisitingPlayers, _
                     Me.VisitingLine)
                End If
            Case "Home"
                If xlr.IsStartElement Then
                    _addPlayersReader(xlr, Me.HomePlayers, _
                     Me.HomeLine)
                End If
        End Select

     Loop
   Catch e As XmlException
       Throw New ApplicationException("Could not load " & fileName, e)
  Finally
      xlr.Close()
  End Try

End Sub
```

KEY POINT

Listing 3-8 is contrasted to Listing 3-7 in that the document is parsed piecemeal, using a `Do` loop and a `Select Case` statement, rather than in a single shot using the `Load` method.[7] This approach has two consequences. First, because the document is being processed incrementally, if the document is not well formed, an `XmlException` will not be thrown until the offending element is reached. Second, even for small XML documents, the `XmlReader` is faster than using the DOM. In fact, even for score sheet documents like these ranging from 4K to 8K in size, the difference is easily measurable with the `XmlReader` being more than 25% faster.

Although this point is not made in Chapter 2, the Compact Framework does not support the `XmlValidatingReader` class that derives from `XmlReader` in the desktop Framework where it can be used to validate an XML

[7] In fact, behind the scenes, the `XmlDocument` class uses an `XmlReader` to parse and load the document into the in-memory tree.

document against a document type definition (DTD), XML-Data Reduced (XDR), or XSD document.

Using XML Writers

The Compact Framework also provides streamed access for writing XML documents by including the `XmlWriter` class. As with `XmlReader`, the `XmlWriter` class is the base class, whereas developers typically work with the `XmlTextWriter` derived class.

Basically, the `XmlTextWriter` includes properties that allow for the control of the XML formatting and namespace usage, methods analogous to other stream writers discussed previously, such as `Flush` and `Close`, and a bevy of `Write` methods that add text to the output stream. An example of writing an XML document using the `XmlTextWriter` is shown in Listing 3-9.

Listing 3–9 *Writing a Document with an `XmlTextWriter`.* This method writes out an XML document representing the box score stored by the `Scoresheet` class.

```
Public Sub WriteXml(ByVal fileName As String)

    Dim fs As FileStream
    Dim tw As XmlTextWriter

    Try
        tw = New XmlTextWriter(fileName, New System.Text.UTF8Encoding)

        tw.Formatting = Formatting.Indented
        tw.Indentation = 4

        ' Write out the header information
        tw.WriteStartDocument()
        tw.WriteComment("Produced on " & Now.ToShortDateString())

        tw.WriteStartElement("BoxScore")
        tw.WriteAttributeString("Visitor", Me.Visitor)
        tw.WriteAttributeString("Home", Me.Home)
        tw.WriteAttributeString("Date", Me.GameDate)
        tw.WriteAttributeString("Time", Me.GameTime)

        ' Visiting team
        tw.WriteStartElement("Visitor")
```

```
        Dim p As PlayerLine
        For Each p In Me.VisitingPlayers
            tw.WriteStartElement("Player")
            tw.WriteAttributeString("Order", p.Order)
            tw.WriteAttributeString("Name", p.Name)
            tw.WriteAttributeString("Position", p.Pos)
            tw.WriteAttributeString("Inning", p.Inning)
            tw.WriteElementString("AB", p.AB)
            tw.WriteElementString("H", p.H)
            ' Other properties here
            tw.WriteEndElement() ' Finish player
        Next

        tw.WriteEndElement() ' Finish visitor

        ' Do the same for the home team

        tw.WriteEndDocument() ' Finish off the document

    Catch e As XmlException
        Throw New ApplicationException("Could not write " & fileName, e)
    Finally
        tw.Close()
    End Try
End Sub
```

You'll notice in Listing 3-9 that the various write methods such as `WriteStartDocument`, `WriteStartElement`, and `WriteAttributeString` are used to write the XML and that the `XmlWriter` is smart enough to close all the open elements with a single call to `WriteEndDocument`. A portion of the resulting XML file follows:

```
<?xml version="1.0" encoding="utf-8"?>
<!-- Produced on 11/9/2002 -->
<BoxScore Visitor="Chicago Cubs" Home="San Francisco Giants"
  Date="08/06/2002" Time="9:15 PM" >
  <Visitor>
      <Player Order="1" Name="Mark Bellhorn" Position="2B"
        Inning="1">
        <AB>4</AB>
        <H>2</H>
      </Player>
  </Visitor>
</BoxScore>
```

Working with Relational Data

As .NET developers are aware, the programming model in the desktop Framework used to manipulate relational data is referred to as ADO.NET and encompasses the classes in the `System.Data` namespace.[8] These classes are substantially supported in the Compact Framework as well and provide a simple and powerful means for developers to work with relational data locally on the device via XML. The primary class in this regard is `DataSet`, which can be used for working with data locally, persisting it on the device, and binding it to controls on a form.

ADO.NET in the Compact Framework

ADO.NET is fundamentally divided between the `DataSet` and its related classes, such as the `DataTable`, and what are called .NET Data Providers. The data set is used to hold a set of relational data in one or more `DataTable` objects, each of which is fully disconnected from the source of its data and, therefore, is ideal for disconnected and occasionally connected mobile scenarios discussed in Chapter 1. A diagram of the `DataSet` class and its child classes can be seen in Figure 3-2.

As a result of the importance of using relational data, the `DataSet` and its associated classes are fully supported in the Compact Framework.

The Compact Framework also ships with two .NET Data Providers, the SqlClient provider in the `System.Data.SqlClient` namespace for accessing a remote SQL Server, as discussed more fully in the next chapter, and SqlServerCe, found in the `System.Data.SqlServerCe` namespace for accessing SQL Server 2000 CE directly on the device, as explicated in Chapter 5. As mentioned in Chapter 2, the Compact Framework does not support the OleDb provider for accessing data. These providers contain all of the classes necessary to communicate with a back-end data store in order to execute commands and retrieve results. For example, the providers support data readers to provide a connected, forward-only, read-only, cursor-style data-access method, which will be more fully discussed in the next chapter. In addition, each provider relies on base classes and interfaces (such as `IDbConnection` and `DbDataAdapter`) found in the `System.Data` and `System.Data.Common` namespaces to provide consistency and polymorphism to developers.

---

[8] For an in-depth look at ADO.NET, see Dan Fox's *Teach Yourself ADO.NET in 21 Days* (Sams, June 2002).

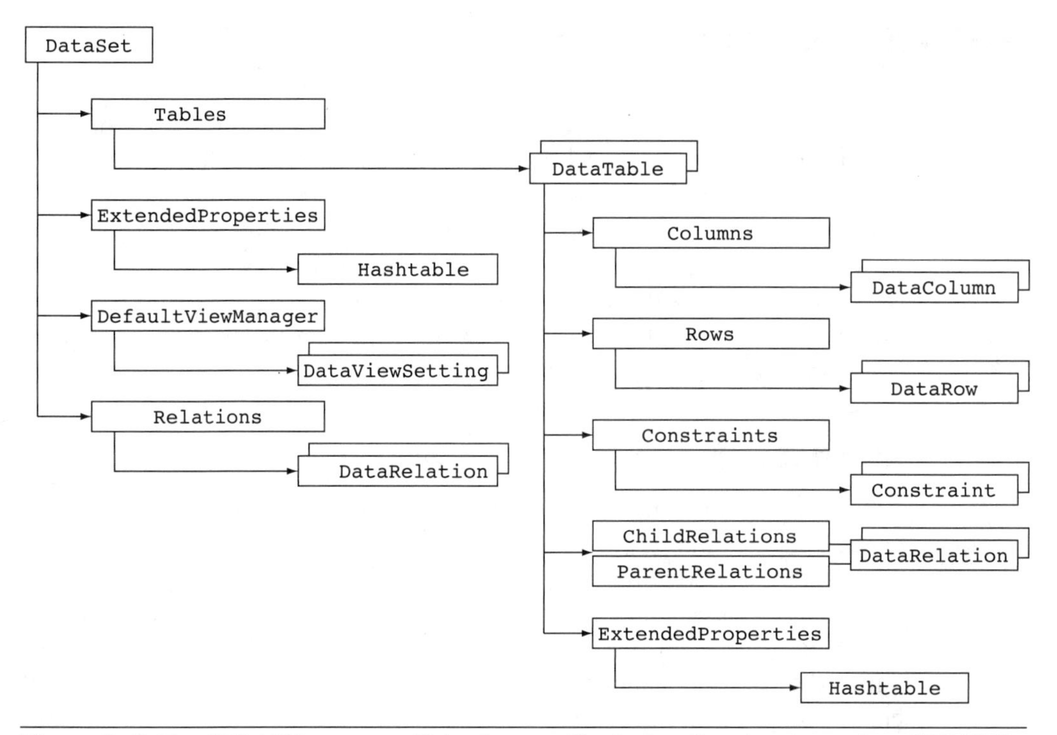

Figure 3–2 *The ADO.NET* `DataSet`. This diagram illustrates the structure of an ADO.NET `DataSet`.

.NET Data Providers, through their data adapters, provide the functionality for retrieving data into, and updating data from, a `DataSet`. In this way, the `DataSet` can be used to work with both remote and local data, and, in fact, the same `DataSet` object can hold data from multiple data sources simultaneously, as shown in Figure 3-3.

Reading and Writing Data

Once data has been populated in a `DataSet` by using a data adapter, that is, an XML Web Service, or even by programmatically loading the `DataSet`, it can easily be saved and reloaded on the device. To save the `DataSet` contents, developers can use the `WriteXml` method, passing it either the filename or the `XmlWriter` to which to write the contents. For example, to write the contents of the `DataSet` to a file, the developer can simply execute the following statement:

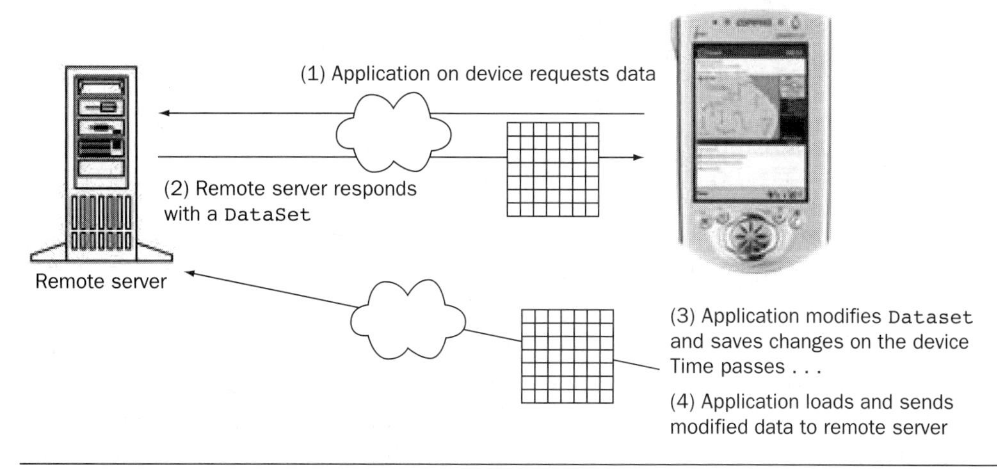
(1) Application on device requests data
(2) Remote server responds with a DataSet
Remote server
(3) Application modifies Dataset and saves changes on the device
Time passes . . .
(4) Application loads and sends modified data to remote server

Figure 3–3 *The Role of the `DataSet`.* This diagram illustrates the use of an ADO.NET `DataSet` in a mobile application. Note that the remote server can be called in a variety of ways, as discussed in Chapter 4.

```
Dim ds As New DataSet()
'Populate the DataSet
ds.WriteXml(fileName)
```

However, it is important to note that `WriteXml` used in this way will create an XML document that contains only the current values in each column of each row in its `DataTable` objects. In other words, although the individual `DataTable` objects can track three different versions of each column value (proposed, original, and current) and four different row states (added, deleted, modified, and unchanged), the XML produced by the previous snippet will not reflect this richness. This is the case regardless of whether the `AcceptChanges` method of the `DataSet` or `DataTable` has been called.[9] As a result, this approach would be useful only when read-only data is being stored on the device, for example, a `DataSet` that contains lookup data later used to populate dropdown list controls and list boxes on a form.

[9] This method sets the `DataSet` or `DataTable` as if all the current values are the original values and all the rows are unchanged. This method is used to provide a clean slate upon which to continue editing data in a `DataSet`.

A better approach that can be used to preserve the changes made to a data set involves using the overloaded signature of `WriteXml` and passing in an `XmlTextWriter`, as follows:

```
Dim xlr As New XmlTextWriter(fileName, System.Text.Encoding.Default)
ds.WriteXml(xlr, XmlWriteMode.DiffGram)
xlr.Close()
```

In this case, the `WriteXml` method accepts both the `XmlTextWriter` and a value from the `XmlWriteMode` enumerated type. Using the value `DiffGram` instructs the `DataSet` to write its contents in the Microsoft Diff-Gram format,[10] which includes both the unchanged data, as well as the before and after versions of any changed values. For example, a very simple `DataSet` that contains just one table with Name and Position columns would look as follows when saved as a DiffGram after changing the position of one of the players:

```
<diffgr:diffgram xmlns:msdata="urn:schemas-microsoft-com:xml-msdata"
  xmlns:diffgr="urn:schemas-microsoft-com:xml-diffgram-v1">
  <NewDataSet>
    <Table diffgr:id="Table1" msdata:roworder="0"
       diffgr:hasChanges="modified">
      <Name>Sammy Sosa</Name>
      <Position>RF</Position>
    </Table>
    <Table diffgr:id="Table1" msdata:roworder="1" >
      <Name>Mark Bellhorn</Name>
      <Position>2B</Position>
    </Table>
  </NewDataSet>
  <diffgr:before>
    <Table diffgr:id="Table1" msdata:roworder="1" >
       <Position>SS</Position>
    </Table>
  </diffgr:before>
</diffgr:diffgram>
```

[10] This format was first introduced as a Web update to SQL Server 2000 to extend SQL Server's XML processing capabilities.

It should also be noted that if a developer wishes to create a DiffGram with only the changed rows and values, he or she can call the `GetChanges` method of the `DataSet` first, creating a `DataSet` with only the modified data prior to `WriteXml`.

To load XML into a `DataSet`, developers can use the `ReadXml` method and pass it either a filename or an `XmlReader`. In either case, if the XML was produced with the `WriteXml` method discussed previously, even if the `DataSet` does not already contain the appropriate schema, the schema will be created automatically. If the XML was not produced with `WriteXml`, the schema is inferred according to rules documented in the VS .NET 2003 online help and may throw an exception.

However, if the file or `XmlReader` contains a DiffGram like that previously shown, the `DataSet` must already contain the appropriate schema, or the schema must be included in the XML document being read. As a result, the technique that developers can use to persist a `DataSet` on a device while preserving its changes is illustrated in the following snippet:

```
Dim xtrData As New XmlTextWriter(fileName,
  System.Text.Encoding.Default)
ds.WriteXml(xtrData, XmlWriteMode.DiffGram And
  XmlWriteMode.WriteSchema)
xtrData.Close()

' Close the app and come back later; ds is a new DataSet

ds.ReadXml(fileName)
```

In this snippet the `DataSet` (`ds`) is saved as a DiffGram as discussed previously. However, the `WriteSchema` value of the `XmlWriteMode` enumerated type is also passed in order to write out the XSD along with the DiffGram. Then, when the application is reloaded, it can simply use the `ReadXml` method to read in both the schema and the DiffGram.

NOTE: The `DataSet` also supports both the `WriteXmlSchema` and `ReadXmlSchema` methods in order to write and read schema information without the associated data.

Displaying Data

Just as in a Windows Forms application in the desktop Framework, the controls in the `System.Windows.Forms` namespace in the Compact Framework can be used to display richly file, XML, or relational data on the device. This can be done using both data binding and manual binding.

Data Binding

KEY POINT

As in the desktop Framework, the forms programming model of the Compact Framework supports data binding through the use of `PropertyManager` and `CurrencyManager` objects exposed through the `BindingContext` property of the `Control` class from which `Form` is derived. In this architecture, any class that implements the `IList` interface[11] can be bound to single and multivalue controls, such as the `TextBox` and `ListBox`, respectively. This includes not only the `DataTable` and `DataView` classes in ADO.NET but also arrays, collections, and even strongly typed collection classes.

Simple Binding

To use simple binding, a developer need only add `Binding` objects to the `DataBindings` collection exposed by a control. For example, in order to bind a `Label` control and several `TextBox` controls to columns in a `DataTable`, execute the following code (where `dt` is the `DataTable`):

```
lblName.DataBindings.Add("Text", dt, "Name")
lblPos.DataBindings.Add("Text", dt, "Position")
txtAB.DataBindings.Add("Text", dt, "AB")
```

In this case, each column is bound to the `Text` property of the respective control. Alternatively, the simple controls could use a `DataGrid`, `ListBox`, or other complex control as the data source so that, if the data source of the complex control changes, the simple controls will also be changed.

When these bindings are created, the form creates a `CurrencyManager` object based on the data source specified and places it in the `BindingContext`

---

[11] This is typically accomplished by a data source by implementing the `IBindingList` interface, which itself implements the `IList` interface and supports additional features such as notification through the `ListChanged` event.

collection. The `CurrencyManager` abstracts the position of the data source and is responsible for updating the controls as the position changes. In this way, a form may contain multiple `CurrencyManager` objects, all independently positioning the same data source. To change the position programmatically, a developer can then simply use the `Position` property of the `CurrencyManager`:

```
Me.BindingContext(dt).Position += 1
```

On simple controls such as the `TextBox`, the contents of the control can be validated using the `Validated` and `Validating` methods. These methods will fire when the `CausesValidation` property of the control that is navigated to (not the control being validated) is set to True. The `Validating` event is where developers put their validation code, while the `Validated` event fires only if the validation succeeds (the `Validating` event returns False in its `CancelEventArgs` parameter) and can be used to reset error messages. Unfortunately, because the Compact Framework does not include the `ErrorProvider` control, developers will need to create their own custom error displays.

NOTE: Of course, deriving from the standard controls such as `TextBox` to create controls that accept only specific input, as shown in Chapter 2, is an even better approach because it alleviates the user from responding to error messages.

Complex Binding

Controls that can display multiple sets of information often expose a `DataSource` property and, therefore, support complex binding. In the Compact Framework, this includes the `ListBox`, `ComboBox`, and `DataGrid` controls.

The `ListBox` and `ComboBox`, which display only a single value at a time, also expose the `DataMember` and `ValueMember` properties to specify the member to display and the value of the displayed member, respectively. For example, to bind a table of a `DataSet` to a `ComboBox` control, a developer would do the following:

```
cbPlayers.DataSource = ds.Tables(0)
cbPlayers.DisplayMember = "Name"
cbPlayers.ValueMember = "Number"
```

In this case the `Name` column of the `DataTable` will be displayed to the user and returned through the `SelectedText` property, while the `Number` column will be tracked and returned using the `SelectedValue` property. In order to allow the control to display the current player for a particular row in a `DataTable` (for example, `dt`, as shown previously), the `SelectedValue` property of the control can be bound using the `DataBindings` collection:

```
cbPlayers.DataBindings.Add("SelectedValue", dt, "Number")
```

KEY POINT

The `DataGrid` control is more interesting because it can display more information for each set of data it is bound to and need only have its `DataSource` property set, as shown in Figure 3-4. However, unlike in the desktop Framework, the Compact Framework `DataGrid` control does not include a drilldown option for displaying detail rows in a child table linked to a master table through a `DataRelation`. In addition, the grid is not edit-

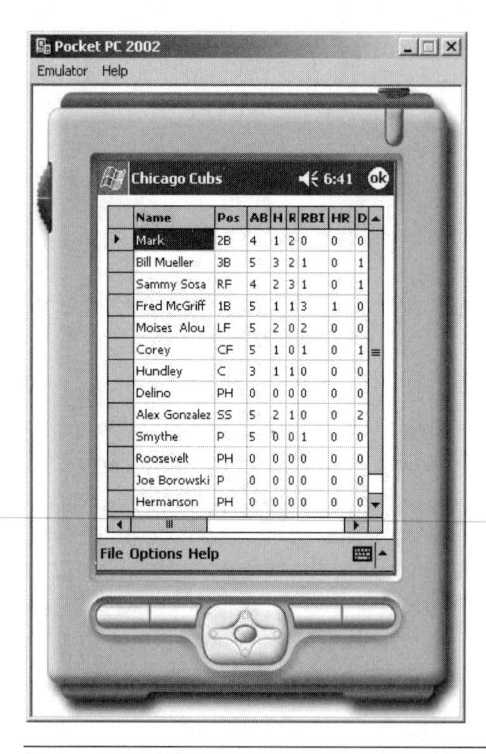

Figure 3–4 *Displaying a `DataGrid`.* This screen shot shows a `DataGrid` bound to a `DataTable` displaying box score information.

able and would be difficult to edit in any case on a small device. As a result, a typical strategy employed by developers is to allow a user to tap a row on a `DataGrid` and then display the detail data on a separate area of the form or more likely on a second form.

The latter technique can be easily accomplished by creating a second form with the appropriate controls to edit the current row. These controls can then be bound to the same data source (available globally within the application) as the `DataGrid` discussed previously. The key, however, is to set the `BindingContext` property of the form with the detail controls to the `BindingContext` of the form containing the `DataGrid`. In this way the detail form will always show the current row in the `DataGrid`. This code can then be placed in the `Click` event of the `DataGrid`, as follows:

```
Private Sub myGrid_Click(ByVal sender As System.Object, _
  ByVal e As System.EventArgs) Handles myGrid.Click

    Dim f As New frmDetail()
    f.BindingContext = Me.BindingContext
    f.ShowDialog()

End Sub
```

In this way, the developer needn't write any code to synchronize the `CurrencyManager` objects on the respective forms or write any manual binding code.

TIP: On the detail form, a developer will want to use the `InputPanel` control to allow the user to use the SIP to enter information into the controls.

Strongly Typed Collections

Complex data binding can be used not only with ADO.NET (`DataTable`, `DataView`, or `DataViewManager`), but with any object that supports the `IList` interface. This includes arrays and other collection classes such as the `ArrayList`. For example, the `ArrayList` exposed by the `Visiting-Players` field in the `Scoresheet` class shown previously can be bound directly to a `DataGrid`, as shown in this snippet:

```
myGrid.DataSource = s.VisitingPlayers
```

In this case the public properties of the `PlayerLine` objects stored in the `ArrayList` are bound to the grid. Public fields will not be bound. The only other caveat is that all of the objects in the `ArrayList` must be of the same type.[12] In order to ensure that this is the case, a developer can create a strongly typed collection class that accepts only objects of a certain type. The advantages to this approach are that developers can work with objects and collections, which is more straightforward than using ADO.NET, and they can take advantage of IntelliSense in VS .NET. For more information on how this is done in .NET, see the article on www.Builder.com in the "Related Reading" section of this chapter.

Manual Binding

Of course, some controls do not support complex binding. This is the case, for example, with the `ListView` and `TreeView` controls that developers often use to display hierarchical data. In order to use these controls with data, a developer must manually bind them. For example, the code in Listing 3-10 displays box score data contained in an `ArrayList` in a `ListView` control passed into the method.

Listing 3–10 *Manual Binding*. This method shows how a developer can manually bind an `ArrayList` to a `ListView` control.

```
Private Sub _showBoxScore(ByVal team As ArrayList, ByVal lv As
  ListView)

    Dim p As PlayerLine
    Dim o As Object

    For Each o In team
        p = DirectCast(o, PlayerLine)
        Dim lvItem As New ListViewItem(p.Name)
         With lvItem
            .SubItems.Add(p.AB)
             .SubItems.Add(p.R)
            .SubItems.Add(p.H)
            .SubItems.Add(p.RBI)
            .SubItems.Add(p.D)
```

[12] Actually, they all must be of the type of the first object in the collection or of a type derived from that type. If not, the row will display, but it will contain an "X."

```
            .SubItems.Add(p.T)
            .SubItems.Add(p.HR)
            .SubItems.Add(p.K)
            .SubItems.Add(p.BB)
            .SubItems.Add(p.LOB)
            .SubItems.Add(p.Pitches)
        End With
        lv.Items.Add(lvItem)
    Next

End Sub
```

Note that because the `ArrayList` contains items of type `Object`; each object in the collection must be cast to `PlayerLine` using the `DirectCast` statement. Subitems are added to each `ListViewItem` to display the properties of the player. A screen shot showing the completed `ListView` is shown in Figure 3-5. Note that each team's box score is displayed on a separate tab

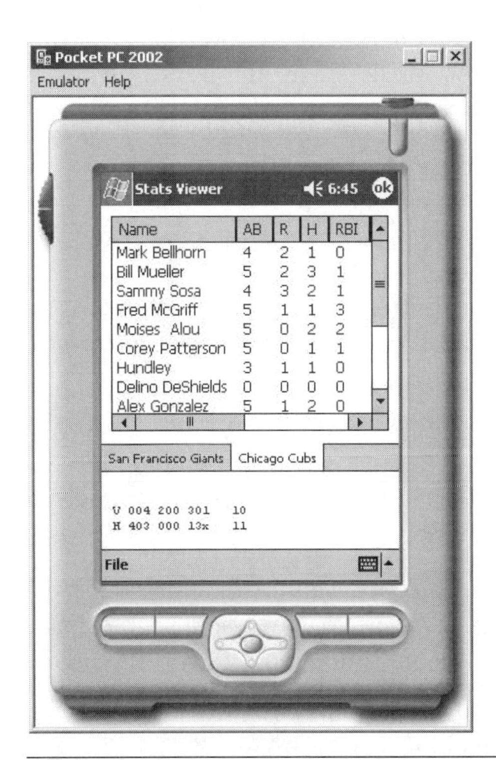

Figure 3–5 *Displaying a `ListView`.* This screen shot shows two `ListView` controls hosted in a tab control used to display box score information.

using the `TabControl`, a very appropriate strategy for smart devices with little screen real estate.

NOTE: From a performance perspective we've found that when using an `ArrayList`, manual binding is faster than data binding. Data binding is further slowed when using the `DisplayMember` property. Of course, your mileage may vary.

What's Ahead

In this chapter you've explored the support in the Compact Framework for working with data locally on the device. This has included the support for working with file, XML, and relational data programmatically and persisting the data to the device, as well as the support for data binding in Compact Framework forms. Together these make up the first essential architectural concept. However, most applications run in occasionally connected or connected modes and, so, must query and retrieve data from remote data sources. In the next chapter you'll explore the ways in which this can be done in the Compact Framework.

Related Reading

Fox, Dan. "Take Advantage of Strongly Typed Collection Classes in .NET." Builder.com (October 25, 2002), at http://builder.com.com/article.jhtml?id=u00220021025dlx01.htm.

Fox, Dan, and Jon Box. "An Introduction to P/Invoke and Marshaling on the Microsoft .NET Compact Framework." *Official Microsoft Smart Devices Developer Community* (March 2003), at http://smartdevices.microsoftdev.com/Learn/Articles/501.aspx.

Fox, Dan, and Jon Box. "Advanced P/Invoke on the Microsoft .NET Compact Framework." *Official Microsoft Smart Devices Developer Community* (March 2003), at http://smartdevices.microsoftdev.com/Learn/Articles/500.aspx.

Fox, Dan. *Teach Yourself ADO.NET in 21 Days*. Sams, 2002. ISBN 0-672-32386-9. See especially Chapters 3–5 and 15.

CHAPTER 4

Accessing Remote Data

Executive Summary

Truly mobile applications go anywhere. While several factors go into the recipe for mobile applications, the ability to access remote data while disconnected or away from the office is an important ingredient. Windows CE provides several mechanisms for accessing data. With the .NET Compact Framework running atop of this Win32 operating system, a developer has additional runtime supports that make various types of communications possible for remote applications.

In the design of mobile applications, several areas should be investigated that include application characteristics, the distance of the device from a network connection, throughput of the available network-connectivity options, communication hardware, and costs. Each of these has multiple considerations that influence how the application should be designed and what Compact Framework technologies are used.

Once these factors are addressed and the project is a "go," the Compact Framework is ready for action. The framework provides different functionalities in the `System.Net` namespace that give a disconnected application the ability to push and/or pull data programmatically from the back end.

At the highest level, the Compact Framework allows an application to be an XML Web Service client (or consumer). This works the same as for desktop Framework clients because VS .NET provides the same experience in the IDE to connect to the Web Service.

Another alternative at a higher level is using the `System.Data.SqlClient` namespace to access SQL Server 7 and 2000 databases. Although easily understood by developers today, this option has several limitations when used from the Compact Framework, including synchronization and scalability.

Yet another mechanism for network communication in the desktop and Compact Frameworks is the availability of Pluggable Protocols. This architecture provides an application a way to work with different protocols without having to code for each one. The only provided class in this scheme for the Compact Framework allows HTTP-based communication. However, the architecture is extensible so that applications can add other protocol support classes as needed.

If lower levels of communication are needed, the Compact Framework includes classes that reside in the `System.Net.Sockets` namespace and are based on Socket Win32 implementation. This list includes the `Socket` class and the socket implementations of TCP and UDP protocols (`TcpClient` and `TcpListener` and `UdpClient`, respectively). Additionally, support for infrared communication exists in the `IrDA` namespace, allowing a developer to communicate easily with other infrared-based devices.

When dealing with occasionally connected devices, the application on the device usually stores data collected from tasks and later synchronizes to a back-end system. A professional feature of such applications is automatic synchronization when connectivity occurs. The Compact Framework does not provide a method to do this, although it can be accomplished programmatically.

Finally, with the myriad ways of accessing remote data, it is important that the UI remain responsive as the application connects and synchronizes data. The Compact Framework gives the developer asynchronous support in `System.Net` so that Compact Framework applications can maintain a responsive UI.

Have PDA, Will Travel

In order to host applications, smart devices require characteristics that include being small and portable, having a decent battery life, displaying information with richness and quality, having memory to hold plenty of information and applications, and the ability to access remote data. This chapter will explore issues associated with the last item in that list. The previous chapter looked at the first essential architectural concept of mobile applications, which is working with data locally. Although this is important, the idea of using mobile applications for business would not exist without the ability to access data that exists on other systems or devices.

Everyone has opened up his or her first PDA like it was Christmas morning. It was easy to enjoy the coolness of being able to carry around a device that contained a considerable amount of information and to use an application that would access it (which demonstrates the first essential concept). But visiting the cradle periodically was required for the device to have the ultimate value of using current information. Accessing remote data in order to provide it to mobile applications is the second essential concept (the other two essential concepts are covered in this scenario as well, but these will be explored in the following chapters) and mirrors the way individuals and organizations have grown in their use of smart devices.

In today's business world, successful companies must become increasingly more agile. This includes being sensitive to changing market factors and understanding changing customer needs. The companies that change more quickly usually live longer and grow wealthier. Therefore, all sorts of information have to be available and accessible anytime and anywhere. Consequently, it is not just a stand-alone PDA environment anymore. The trend is for corporate applications to migrate to mobile devices, and so, it is no surprise that those devices absolutely require the ability to communicate to enterprise back ends.

The integration of mobile devices, the Internet, and wireless connectivity provides a way for companies to extend their reach to employees, partners, and clients. The potential impact of this convergence is unlimited. Possible results include improved productivity, reduced operational costs, and increased satisfaction for all involved. Because of the geographic expansion of wireless connectivity, as discussed in Chapter 1, devices can retrieve information from anywhere in the world.

When looking at potential mobile applications, many factors must be considered. These factors affect not only the cost of operating the application, but also the architecture. It's easy to look at the cost of the device as an expense. But what about other factors? To answer this, factors related to accessing data remotely are explored in the first part of this chapter. After this review, you will have a better idea of the different issues in distributed mobile applications and the types of connectivity provided by today's smart services.

It turns out that the Compact Framework provides a plethora of assistance in accessing data remotely. A Compact Framework–based application can work at different levels of the network stack. This includes TCP/IP, sockets and infrared, streams, HTTP, and XML Web Services. And so, the majority of the chapter will be focused on this support.

Wireless Application Factors around Accessing Remote Data

As mentioned earlier, there are many factors when planning a mobile solution. These factors are mentioned all through this book, but this section will focus on issues related to connectivity. If you are planning your company's first smart device application, this section serves as a starting point.

Application Factors

The first basic issue to research is the location of the data. This should lead to several questions like the following:

- *Must the device remain connected to the back end while using the application?* This scenario is more typical of a manufacturing environment where local access is direct to the company LAN. If required for a device that is outside of the office, then a wireless provider will be sought. Also, the need for real-time data could be different when inside, rather than outside, the firewall and is specific to the application.
- *Can the device connect periodically to the back end?* This is more typical of devices "in the field" in two scenarios. The first is where devices contain information related to a field activity and the information is used as a reference. The second is where a field person collects information that is periodically sent to the back-end system.
- *Is the back-end data located on a machine so that it is accessible from the network or Internet?* Many mobile applications will be enhancements to existing corporate or department applications; therefore, systems planning will be required. While not specific to mobile-application factors, this issue does exist frequently. How many applications do you know about that started as a stand-alone PC application and evolved into some type of multiuser system?
- *Can enough of the business logic from the back end be duplicated on the device in order to make the application and returning data valid?* There are several issues here. First, the processing of a mainframe system cannot be moved to a mobile device (yet) due to memory and processor constraints. Second, the business logic of an existing application should exist in such a form that it can be factored into manageable pieces, some of which can be reused on the device. These issues frequently arises for organizations that have developed so called fat clients that later need to be broken up and their business logic distributed to the middle tier. The existence of good data validation and middle-tier logic make the process of developing a smart device application that much simpler because the logic can be reused rather than rewritten.

Connection Type and Distance

Of these questions, two were related to the location of the data and how often it is updated. Related to these questions is the issue of how far the device is from network. Or in other words, what are the connectivity options based on where the device will be used? The following lists several options:

- *Direct connection:* This is available via the cradle and ActiveSync, which provide Internet access to the device (one might argue that a mobile device connected via a data cable to a cellular phone fits the criteria; however, this belongs to the cellular network category). This might be required by companies that do not trust wireless fidelity (Wi-Fi) or the 802.11x standard, which is discussed below. These types of connections can be made from virtually any wired access point, inside or outside the LAN.
- *IrDA:* Some infrared devices can act as network connections if they are within three feet and have an unobstructed line of sight to the server device. This includes not only some cell phones, but also some wireless access points.
- *Personal area network (PAN):* If the device will be within 30 to 100 feet of the network connection device, smart devices that support PAN connectivity are an option. This is probably known to most as BlueTooth technology, a standard that allows wireless communication between devices at short distances and provides bandwidth up to 1Mbps. One scenario would allow for a device to communicate with a BlueTooth-enabled access point; this would be an in-office scenario. In another scenario, the device would communicate with a cellular phone in order to access the Internet; this would be an in-the-field scenario. In these scenarios, the next critical factor is the bandwidth of the base device. Obviously, an access point would be preferred, because it is connected directly to a traditional LAN (10 to 100 Mbps) instead of a cellular phone, which uses connections that range from 14.4 to 128 Kbps. The PAN standard, also known as WPAN, is managed by the IEEE 802.15 Group.[1] They are working with another group called the BlueTooth Special Interest Group (SIG),[2] which based the BlueTooth standard on work previously done in IEEE activities. This cooperative effort is now known as 802.15.1, where BlueTooth is the PAN standard.

[1] The official Web site for the IEEE 802.15 Working Group is http://grouper.ieee.org/groups/802/11.

[2] The home page of this group is www.bluetooth.com.

- *WLAN:* This standard provides for devices up to one mile away to connect to a base station, otherwise known as an access point, which is the connection to the LAN or Internet. The specification typically includes one of the IEEE 802.11 wireless protocols that provides an 11-Mbps connection. Many corporate environments have embraced this wireless network standard, but corporations are cautious about deploying this technology due to its ease of use by noninvited outsiders. The Wi-Fi Alliance is a group that provides certification of 802.11b-based products. This group provides a certification process so that consumers can have confidence that a purchased wireless product is interoperable with other Wi-Fi products. As of this writing, about 200 member companies have manufactured over 500 different certified products. Therefore, it is a good idea to ensure that purchased products have the Wi-Fi logo.
- *Wide area network (WAN):* WANs are networks provided by the cellular companies and provide reach for mobile devices that span entire continents. There are a variety of network standards and a plethora of service providers; consequently, there is no universal standard. The transfer rates range today from 14.4Kbps to 114Kbps, depending on network type; however, providers are pushing this envelope as data services become mission critical. The vendor choices will be constrained by geographic region; therefore, the vendor's network type will affect the application's throughput and operation cost (because some vendors bill based on time, while others bill on amount of data). The network types will be mentioned throughout this section; however, take a moment to review the types in the Table 4-1. Currently, the technology race is composed of GSM, CDMA, and TDMA, while the older CDPD is widely used in metropolitan areas. GSM has the majority of the world's subscribers in the form of cellular and wireless users. A new standard, GPRS, is growing as it provides GSM networks more users running at higher speeds on the same capacity. In reviewing this industry, acronyms for the next generations of these networks will appear, like those for second generation (2G) and third generation (3G). The good news is that wireless communications is a very competitive global environment and continued improvements are to be expected.

- *Satellite.* As an extended form of WAN solution, satellite solutions do exist. While expensive, some applications require this type of communication. Examples include devastated environments or remote

Table 4–1 WAN Network Types

Network Type	Acronym Meaning	Description
CDPD	Cellular Digital Packet Data	A packet-based network that is usually isolated to a metropolitan area
TDMA	Time Division Multiple Access	A circuit-switched network that has market share in North and South America; used in digital cellular communication; most operators looking to upgrade to GSM
CDMA	Code Division Multiple Access	Developed by QualComm as a competitor to TDMA; a circuit-based network that requires a dial-up connection phase; next generations staring to appear (CDMA2000)
GSM	Global System for Mobile Communication	A circuit-based network; the global leader due to overwhelming market share in Europe and Asia; uses a form of TDMA
GPRS	General Packet Radio Service	A packed-based wireless communication service that has higher data rates and continuous connection to the Internet; implemented on a GSM network; not related to GPS

regions. The speeds are typically less than the 9.6Kbps, and the costs can be much higher than those for typical WAN options.

The various options for connecting smart devices are illustrated in Figure 4-1, with those spanning smaller distances nearer the center of the circle.

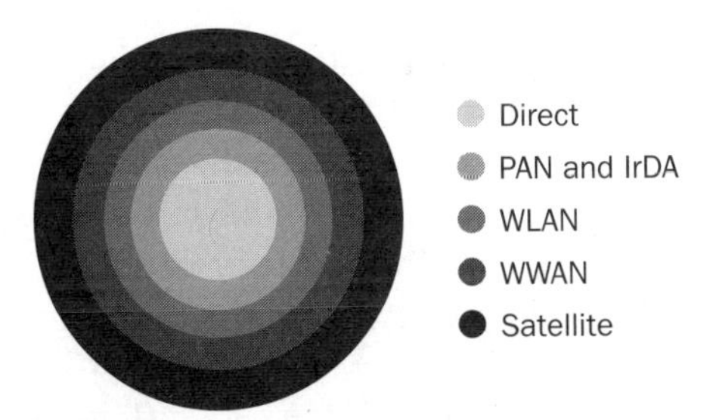

Figure 4–1 *Smart Device Connection Options*. This diagram shows the different types of connections that can be made from mobile devices in order to access remote data.

Throughput

For this discussion, "throughput" is defined as how much data can be moved through a network connection in a given amount of time. For 802.11 connections, this is not an issue at 11 MB. But for others, consider the following questions:

- *For a remote connection, will the available providers' network types have enough throughput?* In scenarios where connections cannot be maintained due to either cost of the call or intermittent connectivity, this will require that data be cached locally and synchronized periodically. If the application requires that data be current, throughput will be a factor, especially when the amount of data is large.
- *For slower connections, will the slower connection last long enough for an acceptable synchronization?* This question raises multiple issues, including throughput, battery power (see the following), intelligent and reliable synchronization, and the reliability of the vendor's network.

Many of these considerations are illustrated in Figure 4-2, which shows the kinds of application scenarios that might be suitable for the various types of connections discussed previously.

Battery Power

As you might expect, battery power is always an issue with mobile devices, and it becomes an even more prominent one when those devices use power for remote connections. As a result, the following basic questions must be asked:

- *Does the attached communication hardware drain the mobile device's battery?* In all scenarios, communications will decrease the device's battery power more quickly than it will when a connection is not active. The difference in power consumption can be quite dramatic and can halve the effective battery life of the device (or worse).
- *Will the attached or integrated communication hardware be used for extended periods due to slower connections?* Although it may seem obvious, this issue has to be considered. However, there are options, including purchasing additional batteries, using chargers in vehicles, and cradling the device (if the application allows).

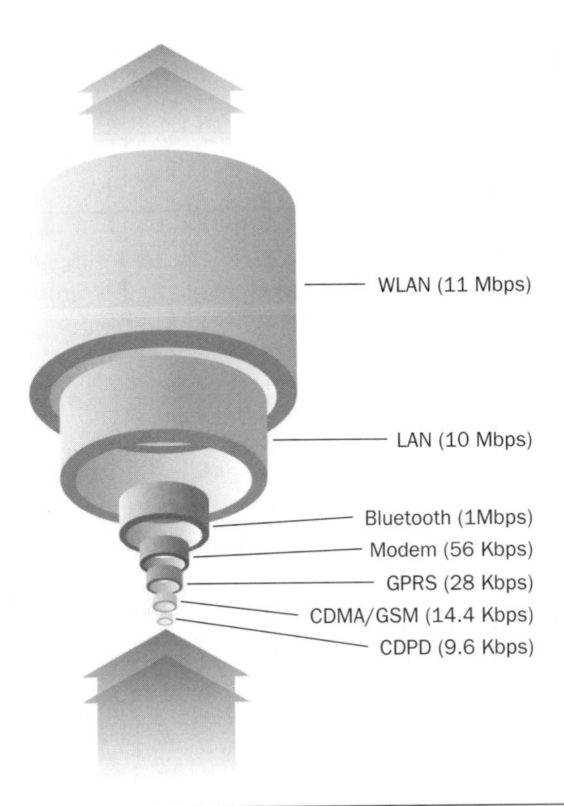

Figure 4–2 *Throughput Scenarios*. This diagram highlights the types of network connections, their typical throughput, and appropriate application scenarios.

Communication Hardware for the Mobile Device

There are different ways that the mobile device will connect to the communications hardware. Each has varying costs and possible usage fees.

- *Integrated:* This is a specialized device where the communication type agrees with the application factors and provider options. Some devices, like the Pocket PC Phone, are integrated with GPRS connectivity and, therefore, work as data or cellular connections. Vendors, including Casio, Intermec, and Symbol, have ruggedized products with CDPD, GSM, or WLAN (802.11) capabilities.
- *Two body:* This scenario requires that a mobile device have some type of communication with another device that serves as the mobile device's communication hardware. Examples include a cell phone

with a data cable, using a BlueTooth-enabled cell phone, an infrared-equipped cell phone, or even the old fashioned modem.

- *Detachable clip-on or plug-in module:* Most mobile devices provide a way to add other hardware. For example, the Pocket PC devices support sleeves that have additional slots for PCMCIA cards or CompactFlash cards. Therefore, a vast number of cards fits into this category. All connectivity types are covered, including WLAN and wireless WAN (WWAN). Some sleeves have the communication hardware integrated like the Novatel CDPD Modem.

Cost

As always costs are a factor when determining how to provide connectivity to a mobile device, and so, the following considerations come into play:

- *Based on the provider type and the amount of data, what is the communication cost?* There are several different billing models for WAN links. Models are based on the amount of data transmitted or the duration of the connection, or they are fixed. Obviously, a fixed model is preferred when large amounts of data need to be transmitted.
- *Have all of the hardware costs been considered?* In some cases, hardware that connects to the device will have to be purchased. If talking about PANs or LANs, access points have to be included as well. If a company has existing mobile hardware that does not have integrated communications, the resulting purchases have to be considered in the project planning.
- *What is the price of developing a software driver?* Although it is not typical for network communications, consider the scenario of communicating with hardware that is not supported in the Compact Framework. For example, if data collection is done on a special hardware device, a driver or an interface to the driver may have to be developed. The cost of this development and its maintenance should be included in the planning.

There are plenty of issues to consider when accessing data remotely. When analyzing the application scenario, start with the application factors. This will usually drive the rest of the issues.

How the Compact Framework Addresses Accessing Remote Data

Just as the desktop Framework does, the Compact Framework provides a variety of network communication mechanisms that access remote data. The existence of these classes is important. As pointed out in the introduction to this chapter, the capabilities provided in this area are imperative. Without them, you can take the word "mobile" out of "mobile application" in a Compact Framework scenario. Table 4-2 highlights the various ways connections can be made and the namespaces in which developers will find the classes necessary to create the connection.

Because the mechanisms exist at different levels of complexity, this section starts at the easier, or higher-level, classes and works down to the lower layers. Another way to position the classes is the dependency relationship. In other words, the higher layers depend on the functionality implemented in the lower layers, as shown in Figure 4-3. The remainder of this section will look at the various ways applications can retrieve remote data.

XML Web Services

As discussed in Chapter 2, Microsoft has helped (with the aid of IBM) to push the idea of XML Web Services into the wider world. This is one of the most hyped concepts surrounding the .NET Framework with good reason. To introduce this subject, the benefits of XML Web Services will be considered, and the important pieces of the XML Web Services architecture will be highlighted. Finally, XML Web Services will be explored in the context of SDP by looking at Compact Framework's support and how VS .NET

Table 4–2 Communication Types and Where They Are Exposed in the Compact Framework

Communication Type	Namespace
XML Web Services	`System.Web.Services`
SQL Server	`System.Data.SqlClient`
Web Client	`System.Net`
Sockets	`System.Net.Sockets`
TCP	`System.Net.Sockets`
UDP	`System.Net.Sockets`
Infrared	`System.Net.Sockets`

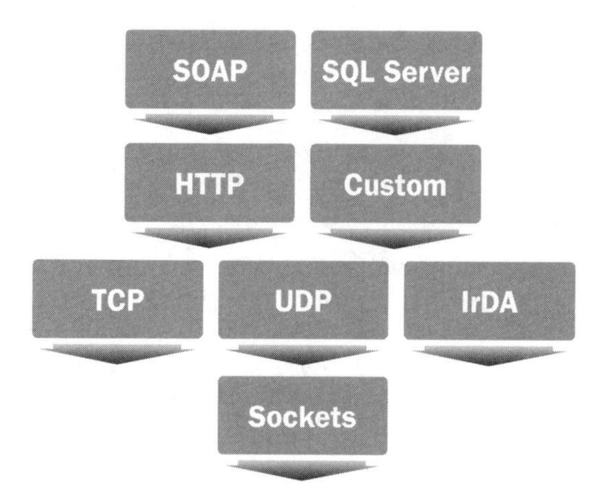

Figure 4–3 *Communication Dependencies.* This diagram positions the communication types in a typical pancake fashion, highlighting the dependencies between the types.

helps the SDP developer. To conclude this section, an example will be reviewed and other related programmatic issues mentioned.

General Benefits of Web Services

By supporting XML Web Services, application-to-application (A2A) communication be can easily achieved. This concept is best illustrated by the platform, language, and device independence discussed in Chapter 2 due to the usage of widely adopted standards like HTTP and XML. However, other benefits exist.

First, one of its biggest benefits is that it makes an enterprise's existing business logic, which may have been tied to a single platform, available to other platforms (internally and externally). In the Windows world, consider the scenario where COM+ objects can now be more easily exposed to internal or external clients for other platforms to use.[3]

Second, the separation of business and data-access logic from presentation logic is always a desirable goal when developing a solution. Moreover, this principle can naturally be applied to application development with Web Services. For example, business logic encapsulated in XML Web Services is naturally partitioned from any presentation logic that accesses it. This natu-

[3] In fact, Windows Server 2003 (COM+ 1.5) includes integrated support for exposing COM+ components, as XML Web Services.

rally allows organizations to develop solutions based on a service-oriented architecture (SOA) model, as proscribed by Microsoft's patterns and practices group.[4] In addition, exposing a Web Service makes it simpler for other applications to retrieve data, rather than having to resort to HTML screen scraping and other fragile techniques.

Third, the communication protocols used with XML Web Services, primarily HTTP, allow Web Service calls to pass through firewalls where other protocols such as DCOM typically have difficulty. Further, emerging standards such as WS-Security[5] allow for secure authentication and encryption when communicating with a Web Service.

Finally, for VS .NET developers, the creation and consumption of XML Web Services are simple tasks that allow the developer to concentrate on application logic instead of the plumbing of Web Services. As will be discussed shortly, both the desktop and Compact Framework provide the classes for handling all of the plumbing issues, including custom authentication, serializing and deserializing the request and the response, and more. Creating a Web Service client is also simple in VS .NET due to its code-generation ability, which will be discussed as well.

Web Services Architecture

An XML Web Service is basically a programmable application component that is accessible via standard Internet protocols. Another way to look at a Web Service is that it is a Web page with no HTML, just data with embedded structure. The architecture of Web Services includes working with a directory of Web Services, performing a discovery of the Web Service, retrieving a description of the Web Service, and then invoking a method on the Web Service, using a known message format.

- *Universal Description Discovery and Integration (UDDI)—the Web Service Directory:* Supported by the likes of Microsoft, IBM, SAP, and other companies that belong to the UDDI Consortium (www.uddi.org), the UDDI specification provides the ability to graphically and programmatically search a database of registered XML Web Services. Replicated directories are hosted by the aforementioned vendors.

---

[4] See the reference to the Microsoft document on the subject in the "Related Readings" section at the end of this chapter.

[5] Microsoft has released the Web Services Extensions (WSE) for the .NET Framework, which implements several of the Global XML Web Services Architecture (GXA) specifications jointly developed by Microsoft and IBM, including WS-Security.

For example, see http://uddi.microsoft.com for Microsoft's hosted directory.

- *DISCO—Discovering the Web Service descriptions:* XML Web Service discovery is the activity of programmatically inquiring for documents that describe the Web Service, including documentation and WSDL (see below). By searching for .disco files on an IIS Web server using a command-line tool or the VS .NET Add Web Reference dialog, clients can discover the endpoints for Web Service on a particular Web server. It should be noted, however, that DISCO is a Microsoft technology that has not been widely adopted and appears to be withering on the vine. For more information about DISCO, see the article referenced in the "Related Reading" section at the end of the chapter.
- *WSDL—Describing the Web Service:* WSDL is an XML grammar that describes the Web Service, including its method names, input parameters and their types, and the response. It is analogous to the metadata found in Compact Framework assemblies and type libraries used in COM. WSDL is an industrywide specification recommended by the W3C.[6] Version 1.1 is implemented by the Framework, although the draft version of 1.2 was made available in March 2003.
- *Invocation and the message format:* SOAP is the XML invocation and message format specification. SOAP is made up of four parts: the requirement of using an envelope to contain the message (the only mandatory requirement), optional rules for encoding and serializing application-specific data, an optional Remote Procedure Call (RPC)-style messaging system (SOAP is really a one-way messaging system), and an optional specification on how to bind the messages to HTTP. Although Web Services are typically invoked via SOAP, they are not required to be. Theoretically, any protocol can be used to transport a Web Service message. However, this limits the use of a Web Service to clients that can utilize the protocol of the hosting platform. So, the main protocols of interest are SOAP, HTTP-GET, and HTTP-POST. Fortunately, VS .NET and the Framework support each of these natively.

[6] The specifications can be found on the Web Services Description Working Group's Web page at www.w3.org/2002/ws/desc.

XML Web Service and SDP

The good news for SDP developers is that the Compact Framework supports the consumption of XML Web Services. Furthermore, the process of creating a Web Reference in an SDP is identical to that in a Windows Forms or ASP.NET application. Because of this similarity, a developer can leverage his existing VS .NET and Framework knowledge while developing an SDP application. This also implies that smart devices will get all of the benefits described in the earlier section on general benefits.

However, as mentioned in Chapter 2, creating XML Web Services is not supported by the Compact Framework because it does not include support for ASP.NET, and Windows CE does not include a Web server.

Creating a Proxy Class for Compact Framework Consumption

In order to call a Web Service without the help of a tool such as VS .NET, a developer would have to go to a great deal of effort. Not only would the developer have to manage XML conversion of the data and messages, he would also have to deal with transmission logic and authentication. Fortunately, VS .NET provides a solution by exposing a code-generation wizard.

VS .NET provides the "Add Web Reference" menu option, which creates a proxy class to encapsulate the invocation of the Web Service. This functionality is purposely analogous to the Add Reference process used to reference code in a local assembly or COM component. In the case of Web Services, however, adding a Web reference indicates the intention of invoking remote code hosted in an XML Web Service.

Once invoked, the resulting dialog shown in Figure 4-4 includes a single step to identify the endpoint (URL) of the Web Service via a known address or searching the UDDI directory. When the host is identified, VS .NET downloads the service's WSDL that describes the Web Service and runs it through a code generator (also exposed in the Wsdl.exe command-line utility) to create a proxy class complete with public methods that map to the service operations and through which the service can be called.

Once built, the proxy class shows up in the Solution Explorer of VS .NET. The class resides in a namespace that reflects the endpoint of the service, although it can be renamed to something more reflective of its function. By examining the proxy class in the Object Browser or Class View window in VS .NET, developers will see that the functionality exposed by the service is now present in the proxy class. In addition to methods used to invoke each operation of the Web Service, the proxy class includes asynchronous function names that start with "`BeginXXX`" and "`EndXXX`," where "`XXX`" is one of the operations exposed by the Web Service.

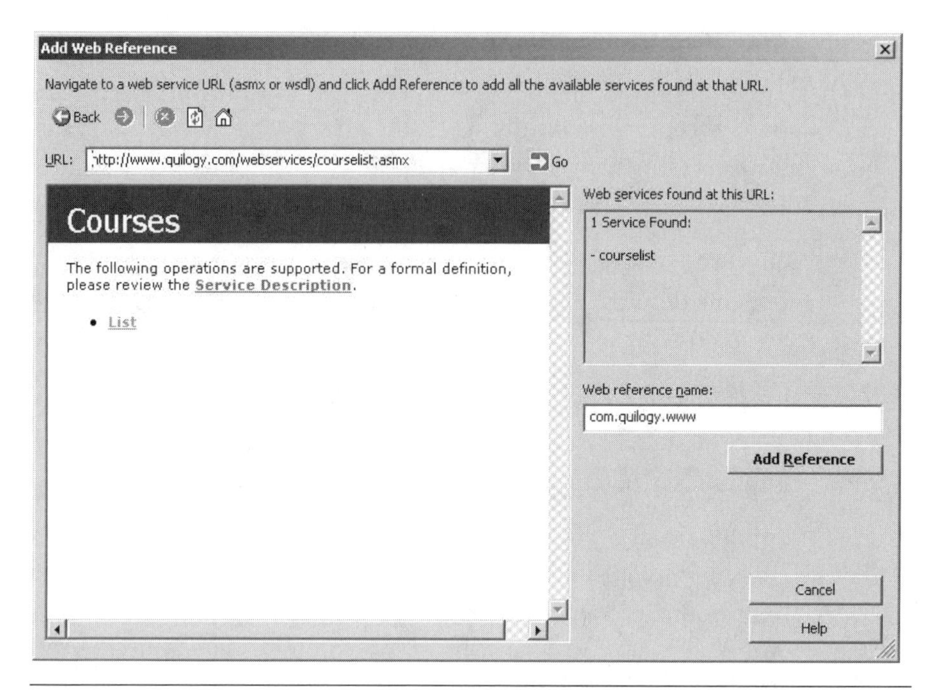

Figure 4–4 *Adding a Web Reference*. This dialog allows a developer to reference a Web Service and downloads the WSDL used to create the proxy class.

Furthermore, the proxy class inherits from the `SoapHttpProtocol` class (and its parents), which includes a core set of properties, including those shown in Table 4-3.

Once the proxy class is built, it is simple to work with the Web Service. The developer would instantiate an object of the proxy class and then call one of the methods. The proxy class handles the rest of the details, including serializing any arguments passed to the service, creating the XML to invoke it, pushing the request to the server hosting the Web Service, and deserializing the response. An additional benefit of using the proxy class is VS .NET's IntelliSense and the compile-time checking against the Web Service method usage. The following code snippet invokes a Web Service that returns an ADO.NET `DataSet` object:

```
' Instantiate proxy object
Dim myService As New ServiceHost.NeededService
  myService.Url = Settings.ServiceEndPoint
' Send request to server and store response in a DataSet
DataSet ds = myService.GetSomeInfo()
```

Table 4–3 Properties Exposed by the Client Proxy Class Used to Call the XML Web Service

Property	Description
`Url`	This property defaults to the value of the endpoint used during creation. It allows the application to override at runtime and is often stored in a configuration file or the registry for ease of maintenance.
`Timeout`	This property defaults to 100,000 ms (100 seconds) and provides a way for the client to specify how long to wait for synchronous method invocations to the Web Service before throwing an exception.
`Credentials`	This property provides for the passing of client credentials to the Web server when the server is configured to use Basic or Windows (IIS only) authentication.
`ClientCertificates`	This property provides for the passing of Authenticode X.509 certificates to the server.
`Proxy`	This property allows for the client to specify credentials for getting through a firewall that is blocking access out to the Internet.

Final Thought on Web Services

If you would like to see some more samples of Compact Framework applications taking advantage of Web Services, there are several good examples out on the Internet with source code, including the following:

- The MapPoint .NET sample[7] application for devices: This application allows users to retrieve interactive maps and driving directions and to find points of interest such as hotels or service stations on their Pocket PCs. MapPoint .NET is Microsoft's first commercially available XML Web Service, providing location intelligence to Web-based, mobile, enterprise, or traditional desktop applications.
- *AskDotNet*[8]: This is an end-to-end sample application for conducting field surveys with devices featuring the .NET Compact Framework, ASP.NET Mobile Controls, and XML Web Services.

Web Services are a natural progression in distributed application architectures. By using the Compact Framework, which supports the consumption

[7] See www.gotdotnet.com/team/netCompact Framework/mappoint.

[8] See www.gotdotnet.com/team/netCompact Framework/askdotnet.

of Web Services, your SDP applications can communicate with a variety of back-end systems.

Accessing SQL Server Remotely

A second way to access data remotely is by using the .NET Data Provider that ships with the Compact Framework and is exposed through the `System.Data.SqlClient` namespace for working with SQL Server.

SQL Server is one of the leading database servers today. As a result, the SqlClient .NET Data Provider is familiar to many desktop and server programmers who work with the .NET Framework because SQL Server is used in Windows Forms, Enterprise Services, and ASP.NET applications today. SQL Server is also well known to those who work with ADO 2.x in VB 6.0 and ASP. Moreover, many developers interested in the Compact Framework will be members of organizations that rely on Microsoft technologies and, therefore, probably have SQL Server databases implemented in important applications. So, a normal transition to smart device architectures will be to access these already established data sources.

As discussed in Chapter 3, ADO.NET is the technology that Framework developers use for accessing data in data sources like database servers. Rooted in the `System.Data` namespace, ADO.NET provides a model that does several things for the developer, including the combination of relational and object-oriented paradigms (OOPs), integration with XML, and the abstraction of database server protocols through .NET Data Providers.

There are several benefits to the way ADO.NET defines data providers. One benefit is that the class model for providers in the desktop Framework is the same as that in the Compact Framework. Therefore, the knowledge of data providers in the .NET Framework can be leveraged except for the actual database implementation differences and smart device constraint considerations. A second benefit is that Compact Framework SqlClient is identical to the desktop Framework implementation of SqlClient. A third benefit is that other data providers that are more easily understood can be built in the Compact Framework. In the case of Compact Framework, the SQL CE team provided a rich data-access mechanism in `System.Data.SqlServerCe`, a data provider for SQL Server 2000 Windows CE Edition (SQLCE) database residing on the device and which will be described in detail in Chapter 5.

ADO.NET Data Providers are required to implement certain interfaces. Therefore, all providers will have managed classes for connections, commands, parameters, data adapters, command builders, data readers, transactions, exceptions, and more. Missing from this is the `DataSet` and its related classes because data sets are provider-independent.

In the remaining parts of this section SqlClient will be explored and considerations related to usage on a constrained device discussed.

The SqlClient Provider Architecture

To get a feel for the SqlClient provider, consider the diagram shown in Figure 4-5. Because of the common .NET Data Provider architecture, certain interfaces must be implemented as depicted in the diagram. Again, this is the same as the SqlClient model for the desktop Framework.

One of the primary advantages of using the SqlClient provider is that it performs very well because it creates Tabular Data Stream (TDS) packets to communicate in SQL Server's native protocol, thereby cutting out any unnecessary layers in the process. This sort of "narrow" provider, which can be used against only one data source, can be differentiated from the "broad" providers, such as those for OLE DB and ODBC, that allow access to multiple data sources and that ship with the desktop, but not with the Compact Framework. Another example of a narrow provider is the Oracle provider that ships with the desktop Framework.

Because many books and articles have been written on ADO.NET and the various aspects of using SqlClient (we shamelessly recommend one of our own books in the "Related Reading" section at the end of the chapter), the discussion will now focus on using SqlClient in SDP applications.

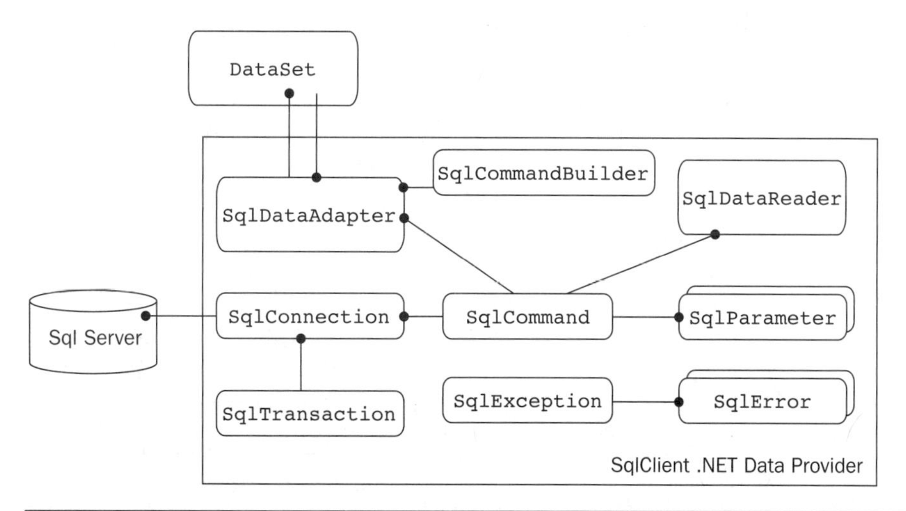

Figure 4–5 *SqlClient .NET Data Provider.* This dialog illustrates the architecture of the SqlClient provider.

SqlClient on a Smart Device

There are several issues to consider when looking at using SqlClient on a smart device.

The first issue to address is whether to use SqlClient or SQLCE. While other chapters will go into more detail on the features of SQLCE, it is important to understand that there are times when accessing a remote SQL Server is a good choice.

For example, consider the following scenario. The device is capturing data in a manufacturing environment, and the data is stored on a SQL Server 2000 database. The business requires the captured data to be placed in the SQL Server database as soon as possible. In most cases, this database exists and is being fed by an existing Web or desktop application. If the device can maintain a full-time high-speed connection to the local network, this scenario works great. We see this in data-collection environments like manufacturing, warehousing, medical, and hospitality.

A second scenario is where the data is too large for the smart device. In this case complex queries can be made to the server that return smaller and more manageable sets of data. This still allows for sending requests to a rich query system provided by SQL Server and also allows for the device to maintain a low-memory footprint. Again, the key requirement for scenarios such as these is a reliable connection to the network, especially in communications with the database.

Although there are scenarios for which SqlClient in a Compact Framework is well suited, there are some limitations to consider:

- The most obvious is that mobility is limited due to the dependency of a full-time, high-speed, reliable network connection.
- For scenarios where devices would access the Internet, the SQL Server has to be exposed to the Internet. Consequently, well-documented firewall ports will have to be opened. However, there are options to change to other ports besides the default (1433).
- Automatic port discovery is not supported. SqlClient has to have the port number defined in the connection string when connecting to a named server instance or not using the default port, 1433.
- In applications where there are periods of no connectivity, some additional programming effort will be required. The downloaded data will have to be cached using a `DataSet`, as shown in Chapter 3. When the time comes for updating the SQL Server, a connection will have to be made and transmission code executed. These are simple tasks, but the plumbing around good synchronization takes additional

effort. Synchronization is not only just the act of exchanging data, but also of providing for error recovery and related issues. This is a primary reason for using SQLCE, which provides built-in synchronization, as discussed in Chapter 7.

- Because the client is connecting directly to the database, scalability of the database is reduced, because this is a two-tier architecture. This can be a factor when a large number of devices is deployed in an environment like a corporate campus or hospital.
- Encryption is not supported. If the database server even requires a certificate, an exception will be thrown. One secure way, however, of connecting remotely to SQL Server is to use a VPN, as discussed in Chapter 9.
- Even though SqlClient in the desktop Framework can communicate over several different protocols (TCP/IP, Named Pipes, AppleTalk, Banyan, NWLink, IPX/SPX, and VIA) due to the underlying Net-Library, SqlClient in the Compact Framework supports only TCP/IP. Therefore, the SQL Server must have this protocol enabled.
- There is no connection pooling on a smart device. This is to be expected because a mobile device is used by one user at a time.

One last and important point regarding SqlClient in the Compact Framework is that even though it does not support collecting credentials using the Windows Security Support Provider Interface (SSPI), it does allow for Windows account credentials to be passed in the connection string, thus not requiring SQL Security to be configured on the server. To do this, the developer has to use the `Integrated Security` parameter, as in the desktop Framework, and also include the `User Id` and `Password` attributes as well, as shown in Listing 4-1.

Listing 4–1 *A Simple Example of Using SqlClient and a* `DataSet`*.*

```
Imports System.Data
Imports System.Data.SqlClient

Private Function GetPublishersDS()
    Dim ds As New DataSet
    Dim strCn as String = "Server=10.0.0.1;Database=Atomic;" & _
      "Integrated Security=SSPI;User Id=STL/PUJOLS;Password=MVP"
    Dim sqlConn As New SqlConnection(strCn)
    sqlConn.Open()
```

```
    Dim sqlDA As New SqlDataAdapter("SELECT * FROM products", sqlConn)

    sqlDA.Fill(ds)
    sqlConn.Close()

    return ds
End Function
```

Using Data Readers

Data readers are an important part of ADO.NET as they provide a fast and efficient way to access data. Simply defined, data readers are mechanisms designed to retrieve forward-only, read-only result sets from a data source. Unlike `DataSet` objects, data readers provide a mechanism to stream through data quickly without caching it for future use. Several of the other differences between data readers and the data sets are summarized in Table 4-4.

In the SqlClient provider the data reader is exposed in the `SqlDataReader` class that implements the `IDataReader` and `IDataRecord` interfaces. `IDataReader` provides important functionality, most notably the `Read` method, which retrieves the next record, while `IDataRecord` provides the methods that allow access to the data in the current record.

Table 4–4 Differences between the `DataSet` and Data Readers

Data Set	Data Reader
Is independent of any provider	Is specific to the provider
Stores heterogeneous data (i.e., from multiple sources)	Retrieves data from a single source
Uses data adapter for updates	Is read-only; uses `Command` object and `ExecuteNonQuery`
Does not require a persistent connection	Requires a connection while open
Can be inherited from using typed `DataSet` objects	Is weakly typed
Performs well, but requires extra resource for storage of the entire result set	Shows best performance; no resource consumption; developer responsible for storage of result set
Can be serialized to XML and passed to and returned from a Web service	Cannot be serialized and passed remotely
Data binding supported	Data binding not supported

Incidentally, the `Read` method of the `IDataReader` interface addresses a common issue that ADO programmers often encounter when using the ADO `Recordset` object. Because the `Read` method both returns False when the result set has been fully processed and moves to the next record, there is no possibility of forgetting to move the cursor, which often happened when the `MoveNext` method of the `RecordSet` was forgotten. An example of opening and traversing a `SqlDataReader` is shown in Listing 4-2.

Listing 4–2 *Retrieving Data Using the SqlDataReader.*

```
Private Function GetTeamsForLibertyBowl() As ArrayList
  Dim al As New ArrayList()
  Dim dr As SqlDataReader
  Dim cn As New SqlConnection(_connection)
  Dim cm As New SqlCommand(teamsSql, cn)

  cn.Open()
  dr = cm.ExecuteReader(CommandBehavior.CloseConnection)

  Do While dr.Read()
    al.Add(New TeamStruct(dr["teamId"], dr["teamName"]))
  Loop

  dr.Close()

  Return al
End Function
```

As mentioned in Table 4-4 and illustrated in Listing 4-2, the lifetime of a data reader is dependent on the lifetime of the connection, whereas the `DataSet` remains even after the connection is closed. In fact, the data adapter classes used to populate `DataSet` objects, such as the `SqlDataAdapter`, use data readers internally to populate the `DataSet`. This is also evident in the fact that a data reader is created only through the `ExecuteReader` and does not provide a public constructor.

Another comparison between `DataSet` objects and data readers is in the area of binding. As discussed in Chapter 3, controls that can be bound in the Compact Framework must implement the `IList` interface or an interface that can return an `IList` interface like `IListSource`. The `DataSet` class does implement `IListSource`, but data reader classes like `SqlDataReader` do not.

Finally, you'll notice that the `CommandBehavior.CloseConnection` argument can be passed to the `ExecuteReader` method. This method ensures that the database connection is closed when the data reader is closed and is essential for passing the data reader around within an application, such as when returning it from a method. Furthermore, the connection object cannot be used for other actions until the `Close` method is called.

To summarize this discussion, a short list of reasons to use a `DataSet` over a data reader include the following:

- Deferred updating of data at the database
- The retrieval of hierarchical data or from multiple sources that will be related
- Local persistence for later retrieval using the `WriteXml` and `ReadXml` methods
- Binding to Window Forms controls

As a result, if these requirements are not important, a data reader would be more appropriate, especially if the following are true:

- Fast access to forward-only read-only data is required.
- The data will be used immediately and not persisted.

Because devices typically operate in an occasionally connected fashion, Compact Framework developers typically use `DataSet` objects more frequently than data readers.

Other Issues to Consider When Using SqlClient for Data Access

Mature practices require that developers utilize defensive programming, especially in areas where the application is transferring data to remote systems. The ADO.NET architecture provides for exception classes specific to the data provider; thus SqlClient includes the `SqlException` class that acts as a container for one or more `SqlError` objects. `SqlException` objects are created when SQL Server encounters an error not on the client side, which would result in standard Framework exceptions.

To handle database errors gracefully, the listings shown in this chapter should include SEH code using a `Try-Catch` block. So, for example, the code in Listing 4-1 could be rewritten as shown in Listing 4-3 to include a `Try` block. Multiple `Catch` blocks exist so errors can be appropriately handled and in this case logged by a function specifically coded for that error. This can be done because not all exception classes are the same. For example, the `SqlException` class can contain multiple errors, which are stored in a collection of `System.Data.SqlClient.SqlError` instances. The last `Catch` block will catch the remaining error types due to the use of the base class `System.Exception`. And last, the `Finally` block will execute whether an error occurs and can be used to clean up database connections and deallocate other resources.

Listing 4–3 *An Example of Using the `Timeout` Property and SEH.*

```
Private Function GetDS() As DataSet
  Dim ds As New DataSet
  Dim cn As SqlConnection
  Dim da As SqlDataAdapter
  Dim strCN As String = "Server=10.0.0.1;Database=Atomic;" & _
  "UID=Musial;PWD=StanTheMan;Connect Timeout=10"

  Try
    cn = New SqlConnection(_connect)
    da = New SqlDataAdapter(teamSql, cn)
    da.Fill(ds)
  Catch ex As SqlException
    LogSqlError("GetDS", ex)
  Catch ex As Exception
    LogError("GetDS", ex)
  Finally
    If (Not cn Is Nothing) Then cn.Close()
  End Try

  Return ds
End Function
```

Also, keep in mind that the `SqlError` classes expose a `Class` property that SQL Server database administrators refer to as "severity level." If an error occurs with a severity level of 20 or higher, the server will close the connection. If the developer is maintaining a global connection object, this will be an issue.

Another note about Listing 4-3 is the controlling of the `Timeout` value in the connection string. The default time-out for the creation of a connection is 20 seconds, but this can be overridden in the connection string. A value of zero implies an infinite time-out and is therefore not recommended. In this example, the default value has been overridden and set to ten seconds.

Final Considerations When Using SqlClient

KEY POINT

SqlClient is a well-known data provider for those who already use SQL Server, and it works well. However, it should be used in the right situations. Considerations include scenarios where there is the need to access data that is larger than the smart device can store or where the data does not require future updating or searching. The scenario would further require the environment to have full-time, reliable, high-speed connectivity. Due to these requirements, the SqlClient provider is not recommended for disconnected or occasionally connected applications.

For these types of applications, there are better alternatives. For example, XML Web Services might be considered a better architecture because the application will gain scalability due to accessing the middle tier. However, this option still lacks built-in synchronization logic. The premier disconnected architecture will utilize SQLCE and the SqlServerCe provider, as discussed in Chapter 7.

Using Pluggable Protocols

For simple interactions with servers, the Compact Framework provides a layered and extensible mechanism called Pluggable Protocols. This is a class model based on the abstract classes in the `System.Net` namespace, `WebRequest` and `WebResponse`. Concrete classes, which inherit from these abstract classes, are registered as supporting a specific protocol.

As Listing 4-4 shows, the `WebRequest.Create` factory method returns the registered concrete class for that protocol. In this example, an `HttpWebRequest` reference is returned because the Uniform Resource Identifier (URI) passed into the `Create` method contains the prefix HTTP. As it turns out, the only pluggable set in Compact Framework is the HTTP pair `HttpWebRequest` and `HttpWebResponse`, which inherit from `WebRequest` and `WebResponse`, respectively. This model allows for the addition of other protocols, like FTP, WebDAV, or NNTP. Although the desktop Framework includes `FileWebRequest` and `FileWebResponse`, these are not supported in Compact Framework. If an unregistered protocol is used, a `NotSupportedException` is generated from the `WebRequest.Create` method.

Listing 4–4 *Calling a Web Server.* This method sends a request to a Web server and displays the raw response in a Message Box.

```
Imports System.Net
Imports System.IO
```

```
Private Sub GetAtomicNews()
   ' Create URI
   Dim uri As New Uri("http://atomic.quilogy.com")

   ' Returns a HttpWebRequest based on URI
   Dim wreq As WebRequest = WebRequest.Create(uri)

 Try
   ' Send request to specified URL
   Dim wresp As WebResponse = wreq.GetResponse()

   ' Set up StreamReader by connecting to WebResponse stream
   Dim sr As New StreamReader(wresp.GetResponseStream())

   ' Read the entire stream into a string
   Dim respmsg As String = sr.ReadToEnd()

   ' Display response
   MsgBox(respmsg)
 Catch e As Exception
   ' Handle errors
 Finally
   sr.Close()
   wresp.Close()
 End Try

End Sub
```

Using the pluggable protocols, developers can build or obtain additional request and response classes to extend their applications to use these protocols. By programming against the base classes, developers can therefore write polymorphic code that is reusable.

Direct Communication with Sockets

Sometimes developers need to communicate with other systems with a low-level protocol. If a developer wanted to work at the lowest level, then sockets-based programming is the place to start. In the simplest terms, a socket is one side of a two-way communication link that exists between two machines on a network. In programming terms, sockets are an abstraction of Internet and other protocols, as shown in Figure 4.6. Providing applications the ability to view network communications as a stream, sockets give

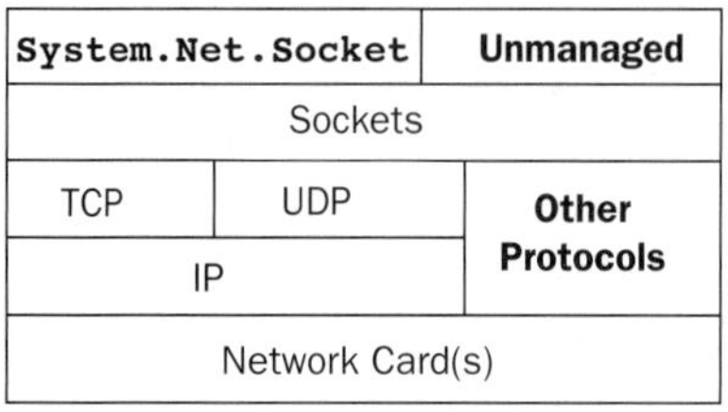

System.Net.Socket		Unmanaged
Sockets		
TCP	UDP	Other Protocols
IP		
Network Card(s)		

Figure 4–6 *Sockets as an Abstraction.* This dialog illustrates that sockets are an abstraction used by developers to access Internet and other protocols.

the ultimate flexibility in communications across different low-level Internet protocols like TCP, UDP, and others. For example, in some enterprise environments, communication with legacy systems can be accomplished using sockets to make requests and return results via TCP or UDP.

The Compact Framework, like its desktop cousin, provides the `System.Net.Socket` class to serve as a wrapper around the native operating system support for networking. All other network-access classes in the `System.Net` namespace are built on top of this managed-socket implementation.

Developers familiar with the PC programming environment know that WinSock is a library that allows socket-based communications for the desktop and server environment. Consequently, Windows CE also has an implementation of WinSock. Just like the PC implementation, WinSock for CE supports the stream and datagram models. The datagram model is used for sending discrete packets to specific addresses; however, this section will focus on the more common stream model.

When working with sockets in a stream scenario, some effort is required to manage a connection. When instantiating a `Socket` object, there are several parts of the constructor, including `AddressType`, `SocketType`, and `ProtocolType`. These are all `System.Net.Sockets` enumerations that provide a variety of networking choices, thus illustrating that the managed-sockets abstraction covers a variety of communication types in the Internet world. As Figure 4-7 shows, the server application has to do a little more work than the client. These steps are typical of any socket library and are shown in Listing 4-5.

Listing 4–5 *A Socket Example of Server Implementation.* This method listens for incoming messages.

```
Imports System.Net
Imports System.IO
```

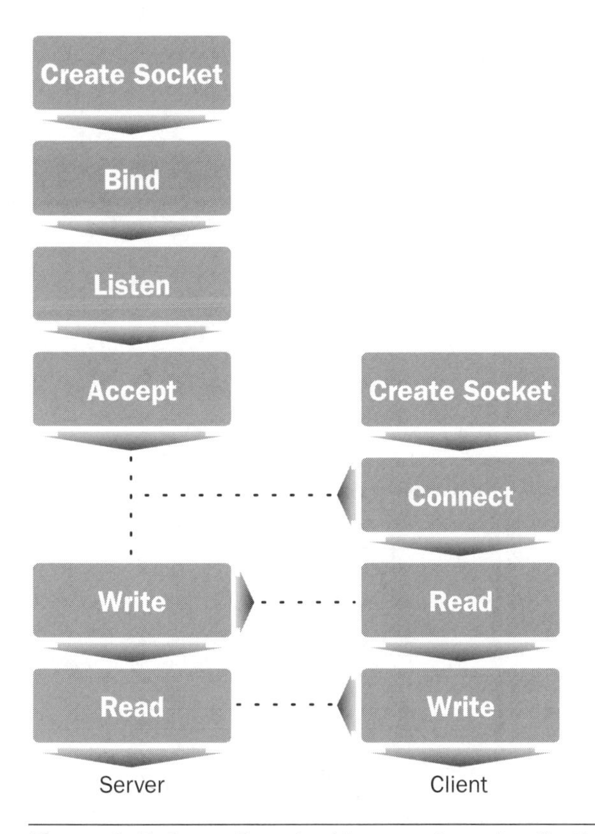

Figure 4–7 *Steps Required in a* Sockets *Application*. This diagram illustrates the steps required of the client and server in order to exchange data using sockets.

```
Public Shared Sub GetSomeData()
    Dim strData As String

    ' Data buffer for incoming data.
    Dim bytes() As Byte = New [Byte](1024) {}

    Dim ipHostInfo As IPHostEntry = Dns.Resolve(Dns.GetHostName())
    Dim ipAddress As IPAddress = ipHostInfo.AddressList(0)
    Dim localEndPoint As New IPEndPoint(ipAddress, 12000)

    ' Create a TCP/IP socket.
    Dim listener As New Socket(AddressFamily.InterNetwork, _
     SocketType.Stream, ProtocolType.Tcp)
```

```
    ' Bind the socket AND listen
    Try
      listener.Bind(localEndPoint)
      listener.Listen(10)

      ' Start listening for connections.
      While True
        Dim handler As Socket = listener.Accept()
        strData = String.Empty
        bytes = New Byte(1024) {}
        Dim bytesRec As Integer = handler.Receive(bytes)
        strData = Encoding.ASCII.GetString(bytes, 0, bytesRec)

        ' Close connection
        handler.Shutdown(SocketShutdown.Both)
        handler.Close()
      End While

    Catch e As Exception
      MsgBox(e.ToString())
    End Try

    If Not listener Is Nothing Then listener.Close()
End Sub
```

The client implementation is easier than the server coding and would look as follows:

```
Dim ipHostInfo As IPHostEntry = Dns.Resolve(hostName)
Dim ipAddress As IPAddress = ipHostInfo.AddressList(0)
Dim ipe As New IPEndPoint(ipAddress, 11000)

Try
  s.Connect(ipe)
Catch se As SocketException
  ' Handle the error
End Try

Dim msg As Byte() = _
  System.Text.Encoding.ASCII.GetBytes("Some data to send")
Dim bytesSent As Integer = s.Send(msg)
```

However, the Compact Framework also includes a higher-level set of classes to work with for doing this type of communication, as the next section discusses.

Communicating with TCP and UDP

TCP and UDP are the most popular Internet protocols today. They are so widely used that the Compact Framework provides managed representations of them. Interestingly, these classes are based on and use the synchronous methods of the `Socket` class and provide an abstraction model that most developers will choose over using the `Socket` class directly.

TCP is more popular than UDP because TCP is responsible for ensuring that data packets are sent to the endpoint and assembled in the correct order when they arrive and because individual UDP messages are limited to 16K. TCP support comes in two flavors: `TcpClient` and `TcpListener`. These classes use the `NetworkStream` class, allowing the developer simply to manipulate the stream using `Read` and `Write` methods. See Listing 4-6 for a `TcpClient` example.

Listing 4–6 *Working with the `TcpClient` Class.* This method sends data to another server.

```
Imports System.Net
Imports System.Net.Sockets

Shared Sub SendMsg(server As String, port as Integer, _
   message As String)

 Dim client As TcpClient
 Try
   client = New TcpClient(server, port)

   'Encode message to ASCII
   Dim data As [Byte]() = _
            System.Text.Encoding.ASCII.GetBytes(message)

   ' Get a client stream for writing.
   Dim stream As NetworkStream = client.GetStream()

   ' Send the message to the TcpServer.
   stream.Write(data, 0, data.Length)
```

```
    data = New [Byte](256) {}

    Dim responseData As [String] = [String].Empty

    ' Read the first of the response bytes.
    Dim bytes As Int32 = stream.Read(data, 0, data.Length)
    responseData = _
             System.Text.Encoding.ASCII.GetString(data, 0, bytes)

  Catch e As SocketException
          ' Log socket error
  Finally
    ' Close local port
    If not client is nothing Then client.Close()
  End Try
End Sub
```

Looking at Listing 4-6, one can see that `TcpClient` usage is straightforward. The constructor defines the local endpoint that expects to get TCP requests. After getting a `NetworkStream` reference from the `GetStream` method, the stream can be filled with data and a response returned before the connection is closed.

Implementing a TCP server requires only a little more effort. First, a `TcpListener` is created and instructed to "listen" to a local port. Upon accepting an incoming message, a `TcpClient` is used, as in the `TcpClient` example, to retrieve the data.

The UDP implementation is included in the `UdpClient` class. Although this class is simple to use due to its `Read` and `Write` methods, this protocol has the negative characteristic of not guaranteeing arrival of packets (or the order of arrival), and therefore, the program has to deal with time-outs, resends, and reassembling of packets. However, one benefit of UDP is that this protocol has some interesting broadcasting services.

Communicating Using Infrared

Infrared ports are included on a variety of devices, including notebook computers and handheld devices. This wireless data-transfer capability is defined by IrDA,[9] which is a group of international companies made up of manufacturers and software companies that promote infrared interoperability.

[9] More information on IrDA can be found at www.irda.org.

The IrDA standards are designed to foster communication on low-cost hardware components that have low power requirements and that easily communicate wirelessly using infrared technology. The purpose of infrared ports is to have a wireless ad hoc data-transfer option, instead of relying on serial cables and serial ports.

IrDA is a short-distance mechanism (up to three feet originally, but results will vary) and was originally designed as half duplex, which means only one endpoint can communicate at a time. The ports should be in a line-of-sight scenario and, therefore, pointed at each other. The IrDA specifications define a protocol stack, which includes connection initialization (handshaking), the frequency of light, device address discovery, data-rate negotiation, information exchange, shutdown, and device address conflict resolution.

In the past, transmission rates have not been blazingly fast, but this story is changing. Implemented under Serial IrDA (SIR or IrDA 1.0), older devices once in the neighborhood of 2.4Kbps now range up to 115.2Kbps half duplex, which is sufficient for simple data transfers. Fast IrDA (FIR or IrDA 1.1) is a specification for full-duplex transfers of up to 4Mbps, which would be acceptable for most applications and is available today. But if high bandwidth is required, keep your eye on Very Fast IrDA (VFIR), which will support up to 16Mbps. Additionally, newer Pocket PC devices such as the iPAQ Pocket PC H3950 also support Consumer IR, which is useful for communicating with consumer electronic devices such as DVD players and televisions and for supporting longer transmission distances.

It is hard to find a Windows CE device that does not have an infrared port. When on a handheld device, there are many application scenarios for infrared data transfer. In most environments, the infrared port can be assigned to a serial port so that control software that understands serial ports only can also support infrared ports.

In the context of programming against these ports, most operating systems provide a socket-based implementation called IrSock, which is built on top of the IrDA stack. The Compact Framework gives a managed interface to this capability. Therefore, terms from the earlier sockets sections will be used, and the code should be recognizable.

Infrared Support in Compact Framework

KEY POINT

The IrDA support in the Compact Framework is for the client side (or the initiating side) of a transfer only. After a developer references the `System.Net.Irda.dll` assembly in the SDP, the `System.Net.Sockets` namespace will have several IrDA-related classes brought into the application. This is one of the few areas where the Compact Framework has functionality that the desktop Framework does not (and if a desktop Framework developer wants this ability, then P/Invoke will be required to call the functionality). The Compact Framework's support for IrDA is based on the Information Access Service (IAS) layer of the IrDA specification. Therefore, the managed implementation comes in three sections of classes: listener, client, and discovery.

As the `RecvIrData` function in Listing 4-7 shows, `IrDAListener` is used to monitor incoming messages. And, just like the earlier socket and `TcpListener` examples, a client class is used to receive the incoming message. The client class is called `IrDAClient`, which maintains a consistent naming pattern. A client application uses `IrDAClient` to send a message to a server application, as shown in the `SendIrData` function.

In order to handle discovery of devices, `IrDAClient` has a method, `DiscoverDevices`, that returns an array of `IrDADeviceInfo` objects, as shown in the `FindAtomicIrDevice` method in Listing 4-7. The `IrDADeviceInfo` object has several properties, including references to other IrDA classes, such as `IrDACharacterSet` and `IrDAHint`. `IrDACharacterSet` is an enumeration that provides for different character sets, including ASCII, Unicode, Arabic, Cyrillic, Greek, Hebrew, and Latin. `IrDAHint` is an enumeration that indicates the device type of a remote device. Example values are computer, fax, printer, modem, telephony, and others.

Listing 4–7 *IrDA Examples.* These methods show the simplicity of working with IrDA.

```
Imports System.Net
Imports System.Net.Sockets

Private ServiceName As String = "ATOMIC_SOCKET"

Private Sub SendIrData(ByVal buffer() As Byte, _
  ByVal bufferLen As Integer)

  Dim client As IrDAClient = Nothing
  Try
    client = New IrDAClient(ServiceName)
  Catch se As SocketException
    ' Handle exception
  End Try

  Dim stream As System.IO.Stream
  Try
```

```
    stream = client.GetStream()
    stream.Write(buffer, 0, bufferLen)
  Finally
    If (Not stream Is Nothing) Then
      stream.Close()
    End If
    If (Not client Is Nothing) Then
      client.Close()
    End If
  End Try
End Sub

''''''''''''''''''''''''''''''''''''''''''''''''''''''''

Private Function RecvIrData(ByVal buffer() As Byte, _
    ByVal bufferLen As Integer) As Integer

  Dim bytesRead As Integer = 0
  Dim listener As IrDAListener = New IrDAListener(ServiceName)
  Dim client As IrDAClient = Nothing
  Dim stream As System.IO.Stream = Nothing

  Try
    listener.Start()

    ' Wait for connection
    client = listener.AcceptIrDAClient()
    stream = client.GetStream()
    bytesRead = stream.Read(buffer, 0, bufferLen)

  Catch e as Exception
    ' Handle the exception

  Finally
    If (Not stream Is Nothing) Then
      stream.Close()
    End If
    If (Not client Is Nothing) Then
      client.Close()
    End If
    listener.Stop()
  End Try

  Return bytesRead
End Function
```

```
''''''''''''''''''''''''''''''''''''''''''''''''''''''''''''''''''''''''''''''
Public Function FindAtomicIrDevice() As IrDADeviceInfo
  Dim irdaClient As New IrDAClient
  Dim irdaDevices() As IrDADeviceInfo
  Dim oDevice As IrDADeviceInfo

  irdaDevices = irdaClient.DiscoverDevices(3)
  For Each oDevice In irdaDevices
    Dim EndPoint As New IrDAEndPoint(oDevice.DeviceID, "IrDA:IrCOMM")
    irdaClient.Connect(EndPoint)
    If oDevice.DeviceName = "AtomicDevice" Then
      Exit For
    End If
  Next
End Function
```

One of the quicker ways to get an IrDA example running is to use the tic-tac-toe example provided with VS .NET 2003. Not only is this well implemented, but the game is not bad either.

Other Issues in Network Communications

Many network capabilities and issues have been discussed in this chapter. In this final section, we will cover two more considerations. First, we will show an important utility function that occasionally connected applications can utilize. Second, we will overview Compact Framework support for asynchronous networking functions.

Checking for a Network Connection without Creating a Connection

In some applications, the smart device will often be disconnected from a network. This architecture will require periodic connections so that synchronization activities can be initiated. Most developers would like the application to sense the connection and perform the synchronization in the background. However, the Compact Framework does not provide any events like this yet.

One scheme for implementing this functionality is to check for the existence of a known Web site on a `timer` event, so that checking for an Internet connection can occur in the background and without the user's knowledge. For cradled device scenarios or devices with a wireless card, this is a satisfactory mechanism. However, this mechanism may not be acceptable for phone-based smart devices, as it will initiate a dial-in to the network every time the code is executed. An application like this would not survive long and would probably cause pain and misery to a support staff.

KEY POINT

Listing 4-8 contains `CheckForNetworkConn`, which is a handy function developers can package in a common assembly and reuse across projects. It is based on what ActiveSync does when the device is cradled. First, the cradle and the PC become a gateway to the network. Second, ActiveSync assigns an IP address to the device. As a result, whenever the device's address is not 127.0.0.1, the device is talking to the network, and in most cases today, that includes the Internet as well. Another benefit of the code is that it works not only for cradled devices, but for Pocket PC Phones as well because it does not initiate the networking dial-in process, which gives the application a little more control. If your organization is building an application that does periodic connections, you'll find this code useful.

Listing 4–8 *Checking for the Existence of a Network Connection.*

```
Imports System.Net

Private Function CheckForNetworkConn() As Boolean
  Dim bRetVal As Boolean = False
  Try
    'returns the Device Name
    Dim HostName As String = Dns.GetHostName()
    Dim thisHost As IPHostEntry = Dns.GetHostByName(HostName)
    Dim thisIpAddr As String = thisHost.AddressList(0).ToString
    bRetVal = thisIpAddr <> IPAddress.Parse("127.0.0.1").ToString()
  Catch ex As Exception
    bRetVal = False
  End Try

  Return bRetVal
End Function
```

Compact Framework Asynchronous Capabilities in Networking

As mentioned in Chapter 3, there are several asynchronous options available in the desktop Framework that are not available in the Compact Framework. Because this chapter is focused on accessing remote data, Table 4-5 shows the communication methods that have Compact Framework–provided

Table 4–5 Asynchronous Support in Network Communication

Namespace	Class	Synchronous Method	Asynchronous Methods
Your application	XML Web Service proxy class	Any WebMethod	Begin[WebMethod name], End[WebMethod name]
System.Net	WebRequest	GetRequestStream	BeginGetRequestStream, EndGetRequestStream
		GetResponse	BeginGetResponse, EndGetResponse
	HttpWebRequest	GetRequestStream	BeginGetRequestStream, EndGetRequestStream
		GetResponse	BeginGetResponse, EndGetResponse
	DNS	Resolve	BeginResolve, EndResolve
		GetHostByName	BeginGetHostByName, EndGetHostByName
	NetworkStream	Read	BeginRead, End Read
		Write	BeginWrite, EndWrite
System.Net.Sockets	Socket	Connect	BeginConnect, EndConnect
		Send	BeginSend, EndSend
		SendTo	BeginSendTo, EndSendTo
		Receive	BeginReceive, EndReceive
		ReceiveFrom	BeginReceiveFrom, EndReceiveFrom
		Accept	BeginAccept, EndAccept

asynchronous support. The list is reflective of the classes covered sequentially in the chapter, except for `DNS` and `NetworkStream`.

One of the Framework design goals is that the calling code should decide if the call is going to be asynchronous or synchronous. This is the reverse of past development environments like COM, where the callee decided how this would work.

In order to understand how to do asynchronous programming, consider an example that would use the asynchronous form of the `GetResponse` method of the `HttpWebRequest` class. Listing 4-9 shows the programming pattern for employing asynchronous methods in which there are typically three steps. The first is to call the appropriate `Begin` method, passing in a delegate that points to a local method called when the method completes. The second is to call the `End` method and retrieve the returned results from inside the local method. The third is to notify the UI that the process has completed and to update the UI.

Listing 4–9 *Using the Asynchronous Version of `HttpWebRequest::GetResponse`.* This sample sends a request to a URL specified in `txtAddress` and returns the response in `txtBody`.

```
Private _inv As Invoker 'from Listing 3-3

'Instantiate _inv in a Form startup event

Private Sub cmdButton1_Click(ByVal sender As System.Object, _
    ByVal e As System.EventArgs) Handles Button1.Click
  txtBody.Text = ""
  Dim sURL As String = txtAddress.Text
  Dim wrGet As HttpWebRequest

  wrGet = CType(WebRequest.Create(sURL), HttpWebRequest)

  Dim myCallback As New AsyncCallback(AddressOf ResponseReceived)
  wrGet.BeginGetResponse(myCallback, wrGet)
End Sub

Private Sub ResponseReceived(ByVal ar As System.IAsyncResult)
  Dim wrGet As HttpWebRequest = CType(ar.AsyncState, HttpWebRequest)
  Dim wrResponse As HttpWebResponse
  Dim responseStream As IO.StreamReader

  wrResponse = CType(wrGet.EndGetResponse(ar), HttpWebResponse)
```

```
  responseStream = New IO.StreamReader(wrResponse.GetResponseStream())

  _inv.Invoke(New UIUpdate(AddressOf SetText), _
     responseStream.ReadToEnd)
End Sub

Private Sub SetText(ByVal oObjects() As Object)
  Me.txtBody.Text = Convert.ToString(oObjects(0))
End Sub
```

Although this is a simple pattern to employ, updating controls on a thread different than the form's thread can lead to exceptions that might be tough to reproduce and solve. The `Form` class `Invoke` method is the solution, although, as discussed in Chapter 3, the `Invoke` method does not allow for the passing of parameters as the desktop Framework method does. The workaround for this was given in the previous chapter through the custom `Invoker` class shown in Listing 3-3.

To see the processing of an asynchronous call in action, developers can do the following:

1. Create a Compact Framework project with two text box controls (named `txtAddress` and `txtBody`) and a button called `cmdButton1`.
2. Insert the code from Listing 4-9 and the `Invoker` class from Listing 3-3.
3. Place breakpoints on the first line of each procedure.
4. Start up the application in debug mode on an emulator or device. Enter a valid URL in the `txtAddress` text box, and click on `cmdButton1`.
5. At this point, you will hit the breakpoint in `cmdButton1_Click`. Enable the Threads windows by selecting Debug | Windows | Threads. Notice the ID column and the value of the one thread, which is the `Form` thread. Press F5 (or the equivalent key, depending on your keyboard layout) to continue debugging.
6. The debugger will stop on the `ResponseReceived` method because the delegate from the `click` event pointed here. Notice in the Thread window that there is a second thread, which is the background thread that executed the `GetResponse`. The code could update the GUI, but this is where the problem could occur. Therefore, the Web server's response is stored using the `Invoker` class `Invoke` method. Continue debugging.
7. The debugger will stop on the `SetText` procedure. Notice that the current thread is the `Form` thread from the first breakpoint. The call

stack is still in the call from the `Invoke` method of the `Invoker` object (which can be proven using the Call Stack window found under the Debug Window menu). Continue debugging to see the raw HTML response.

In addition to the supported asynchronous techniques discussed here, any synchronous functionality can be made asynchronous with the help of the `Thread` class from the `System.Threading` namespace. Refer to the sidebar titled "Synchronizing Access to Resources" in Chapter 3.

With this asynchronous capability, Compact Framework applications can appear responsive, while retrieving and sending data in the background. As always, however, executing tasks in the background does not speed up the program and, if abused, will only slow down the application. So, use asynchronous techniques conservatively and appropriately.

What's Ahead

In this chapter, the discussion focused on the second essential architectural concept: accessing remote data. You've looked at networking considerations for connecting smart devices and also how the Compact Framework supports the ability to send and receive remote data.

In the next chapter, we'll explore the third essential architectural concept, robust data caching using SQLCE.

Related Reading

Fox, Dan. *Teach Yourself ADO.NET in 21 Days*. Sams, 2002. ISBN 0-672-32386-9.

"Microsoft Patterns and Practices," at www.microsoft.com/resources/practices. Specifically, you should note the document titled "Application Architecture for .NET: Designing Applications and Services," at http://msdn.microsoft.com/library/default.asp?url=/library/en-us/dnbda/html/distapp.asp.

IEEE Wireless Standards Zone, at http://standards.ieee.org/wireless/overview.html.

Skonnard, Aaron. "Publishing and Discovering Web Services with DISCO and UDDI." *MSDN Magazine* (February 2002), at http://msdn.microsoft.com/msdnmag/issues/02/02/xml.

CHAPTER 5

Caching Data with SQL Server CE

Executive Summary

Chapter 3 focused on how applications written with the Compact Framework could manipulate data locally on a mobile device. Although this capability is important, many application scenarios require more robust data caching. These scenarios comprise many applications of the occasionally connected type, including sales-force automation, field-service automation, real estate, and home-visit medical applications, among others. For these applications a local relational database that features data integrity, built-in synchronization, access from multiple development environments, and strong security is a must.

Shortly before VS .NET 2003 and the Compact Framework were released, Microsoft shipped SQL Server 2000 Windows CE Edition 2.0 (SQLCE 2.0),[1] which fulfills these requirements. This local database engine consisting of a storage engine and query processor is implemented as an OLE DB provider and runs in-process with Compact Framework applications. SQLCE 2.0 includes a number of new features, including parameterized queries, index seeks, and the `UNION` clause.

To provide access to SQLCE, Compact Framework developers can use the SqlServerCe .NET Data Provider. This provider is implemented using a common set of interfaces and classes and therefore allows developers to leverage their existing ADO.NET knowledge and begin writing applications for SQLCE. The classes in the provider, for example, `SqlCeEngine`, can be used to create databases, tables, and indexes programmatically, in addition to compacting databases and querying and modifying data in disconnected

[1] Some prefer to use the acronym SSCE, but we prefer SQLCE because we believe it is easier to understand.

and connected scenarios. In particular, the ability of SqlServerCe to support index seeks directly on tables can greatly speed the performance of an application because it eliminates the overhead of the query processor. A common approach to manipulating data in a SQLCE database is to write data-access utility classes that encapsulate calls to the provider and distribute the classes to other developers in an organization in an assembly.

Because the Compact Framework supports accessing both a remote SQL Server through the SqlClient .NET Data Provider and local SQLCE databases, developers can use the Abstract Factory design pattern to create factory classes that can be utilized to abstract which provider is used at runtime. This is particularly effective for occasionally connected applications that require access to a remote SQL Server when connected to the network through a WLAN, for example.

SQLCE also offers a high level of security by supporting both password protection and encryption of the database on the device. This is important because the device on which SQLCE is running is inherently mobile and can easily fall into the wrong hands. The encryption algorithm is based on the password; therefore, passwords of eight or more characters are recommended.

SQLCE is included with VS .NET 2003 and automatically installs on the developer workstations. It is deployed automatically when a developer references the SqlServerCe .NET Data Provider in his or her application. It is also possible, and sometimes a preferred strategy, to prebuild SQLCE databases that will be included in RAM or on a storage card and that are large or will be deployed to a large number of devices.

The Role of SQLCE

In Chapter 3 the discussion focused on how to handle file, XML, and relational data locally on a smart device. Part of that discussion showed how data could be persisted on the device and later reretrieved and displayed to the user. Although these techniques are acceptable for some applications, many applications require more sophisticated local storage. To address this need, this chapter focuses on the third essential architectural concept: robust data caching.

History of SQLCE

Along with the release of the Pocket PC 2000 platform, Microsoft anticipated the need to extend the data-management capabilities of enterprise-to-mobile devices. As a result, they shipped SQL Server 2000 Windows CE Edition 1.0 (SQLCE 1.0), code-named "Pegasus," in late 2000 to coincide with the release of SQL Server 2000.[2] SQLCE provides a local relational database of multiple tables running on a device that can be queried with a subset of the Transact-SQL syntax supported in SQL Server 2000. In addition, the first version supported referential integrity, transactions, and even accessing data remotely from SQL Server 6.5, as well as merge replication with SQL Server 2000. Developers could access SQLCE 1.0 using the ActiveX Data Objects for Windows CE 3.1 (ADOCE 3.1) library that shipped with the product.[3] This library, analogous to the ADO 2.x components used with VB 6.0 to access server-based relational data, allowed eVB developers to manipulate SQLCE databases on the device, while eVC++ developers could make OLE DB calls directly using the OLE DB provider for SQL Server CE (OLE DB CE).

Although SQLCE was well received, in order to access remote servers in version 1.0, the device had to be connected to the network via a modem or network card. Microsoft released version 1.1 in June of 2001 and added the SQL Server CE Relay product, which allowed the device to access remote servers when cradled using ActiveSync 3.1. In addition, version 1.1 was included in Microsoft Platform Builder 3.0 as a component and could be deployed as part of an embedded device, as described in Chapter 1.

With the planned release of the .NET Compact Framework in early 2003, it became essential that Compact Framework applications be able to take advantage of SQLCE easily. In September of 2002 Microsoft officially released SQL Server 2000 Windows CE Edition 2.0 (SQLCE 2.0, or from here on out simply SQLCE), which added integration with the Compact Framework through the `System.Data.SqlServerCe` namespace and also added important new functionality in the query processor and storage engine. As a result, applications built on the Compact Framework can use SQLCE for robust data caching directly from managed code and needn't use ADOCE or the OLE DB CE API.

[2] Prior to this, developers had to rely on proprietary schemes, Pocket Access, or the Windows CE database (CEDB) format accessible through Windows CE APIs or through ADOCE, which, to say the least, lacked many of the features developers expect in a relational database system.

[3] In addition, Microsoft released the ActiveX Data Object Extensions for Data Definition Language and Security (ADOXCE) to extend ADOCE for creating, deleting, and modifying schema objects.

NOTE: As you might imagine, because SQLCE and the Compact Framework are separate products, they do not always overlap. For example, SQLCE can be run on Handheld PC 2000 (H/PC 2K) devices, including the HP Jornada 720 and Intermec 6651, and embedded devices built with the Platform Builder 3.0, such as the Intermec 5020. For using SQLCE on these devices, developers will need to use eVB and eVC++.

Robust Data Caching

So what does the term "robust data caching" actually mean? This concept addresses several key elements, including the following:

- *Local relational database access:* SQLCE supports a programming model on a smart device, which most developers are familiar with on the desktop. This allows developers to leverage their skills when creating applications and organizations to extend their data-management capabilities to devices. This is particularly the case for SQL Server developers because SQLCE supports familiar Transact-SQL syntax. Desktop Framework developers will also be able to get up to speed quickly because access to SQLCE is provided through a .NET Data Provider, using the standard classes and interfaces of ADO.NET.
- *Disconnected data integrity:* SQLCE provides a robust cache for storing data on the device for use when the device is disconnected from the network. Because SQLCE supports relational database features such as unique indexes and foreign keys, the data can be manipulated on the device while it is disconnected and still maintain its integrity.
- *Built-in synchronization:* SQLCE includes two different synchronization mechanisms that are built in, thereby saving developers from having to write complex infrastructure code for occasionally connected applications. These techniques will be addressed in Chapter 7.
- *Managed or unmanaged access:* SQLCE is architected as an OLE DB CE provider and, therefore, can be accessed using either the native SqlServerCe .NET Data Provider in the Compact Framework or ADOCE; therefore it can be used with both VS .NET 2003 and eVB/eVC++.
- *Security:* A key part of being a robust data cache is securing the data. As will be discussed later in this chapter, SQLCE supports password protection and data encryption so that data stored on the device is secure.

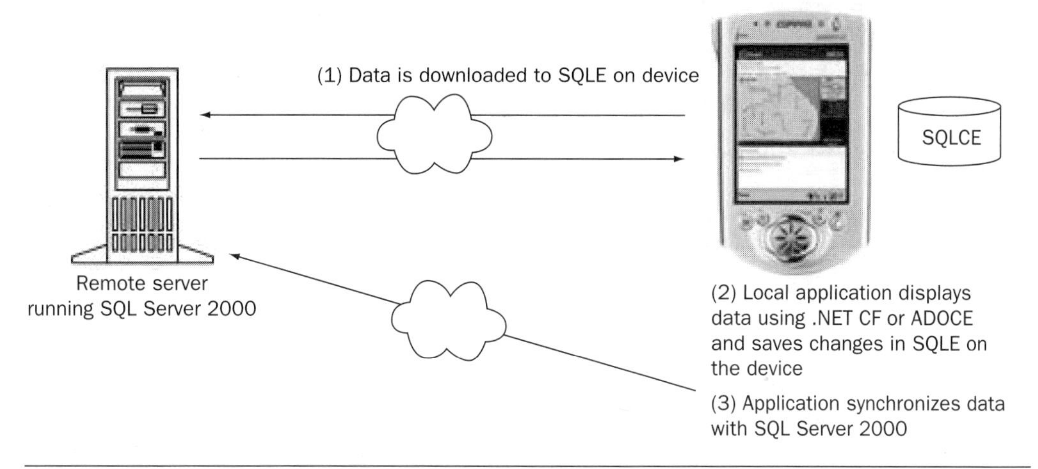
(1) Data is downloaded to SQLE on device
SQLCE
Remote server
running SQL Server 2000
(2) Local application displays
data using .NET CF or ADOCE
and saves changes in SQLE on
the device
(3) Application synchronizes data
with SQL Server 2000

Figure 5–1 *The Role of SQLCE.* This diagram shows the role of SQLCE in storing local copies of data that are then modified on the device and later synchronized with the remote server.

Obviously, applications that can utilize these features run the gamut, but typically they fall into the occasionally connected scenario where data is downloaded to the device, stored in SQLCE, accessed from SQLCE and updated on the device, and then later synchronized with a back-end data store such as SQL Server 2000, as shown in Figure 5-1.

KEY POINT

As you can imagine, an architecture like that shown in Figure 5-1 is useful in a variety of application scenarios, including sales-force automation, where mobile users are downloading and updating customer information; field-service automation, where delivery drivers and maintenance workers download and process deliveries and work orders; real estate, where agents download MLS listings; and medical applications, where doctors and nurses download and update patient and prescription information. Many of these solution scenarios have been implemented and documented as cases studies on the SQL Server CE Web site referenced in the "Related Reading" section at the end of the chapter.

Differences with Local Data Handling

As discussed in Chapter 3, Compact Framework developers can use the ADO.NET `DataSet` object to persist data on the device.[4] As covered in

[4] Although not mentioned in Chapter 3, the Compact Framework does not support typed `DataSet` objects, which are classes derived from the `DataSet` class and generated through a visual designer and code generator in VS .NET.

Chapter 4, the `DataSet` can be populated from a remote server by calling an XML Web Service or by connecting to a remote SQL Server directly, using the SqlClient .NET Data Provider. While these techniques are well suited to some applications, there are several important advantages that a robust data-caching product such as SQLCE provides that ADO.NET cannot:

- *Secure and efficient data storage:* As mentioned previously, SQLCE supports encryption and database passwords. Data stored in `DataSet` objects is persisted to XML and has no such protection. In addition, XML is a verbose format and consumes more memory on the device than does data stored natively in SQLCE.
- *Query access to multiple tables:* Although the `DataSet` object can include multiple `DataTable` objects and can even include primary and foreign keys, it does not provide the ability to query those tables using `SQL` and `JOIN` clauses. SQLCE, as a relational database, makes it easy to query multiple tables through joins.
- *Query performance for large data sets:* As a corollary to the previous point, `DataSet` objects must be programmatically manipulated and, therefore, are much more processor-intensive to query. SQLCE is well suited to querying large amounts of data using a data reader, whereas `DataSet` objects are useful when manipulating between 10 and 100 rows.
- *Performance when populating:* Populating a `DataSet` from an XML Web Service does not offer any support for compression during the transmission of the data, whereas SQLCE does compress data when using its synchronization techniques. As a result, a larger amount of bandwidth is required for accessing the same amount of data with a `DataSet` and XML Web Service.

SQLCE Architecture

Once you understand the role that SQLCE plays in a solution, it is important to understand its architecture and features. The diagram shown in Figure 5-2 illustrates the architecture of SQLCE and the various software components that make up a solution that uses it.

As you can see from Figure 5-2, SQLCE itself is implemented as a DLL and an OLE DB provider (OLE DB CE) that can be accessed from both managed code using the .NET Data Provider for SQL Server CE (SqlServerCe is discussed in the following section) and eVB using ADOCE and directly

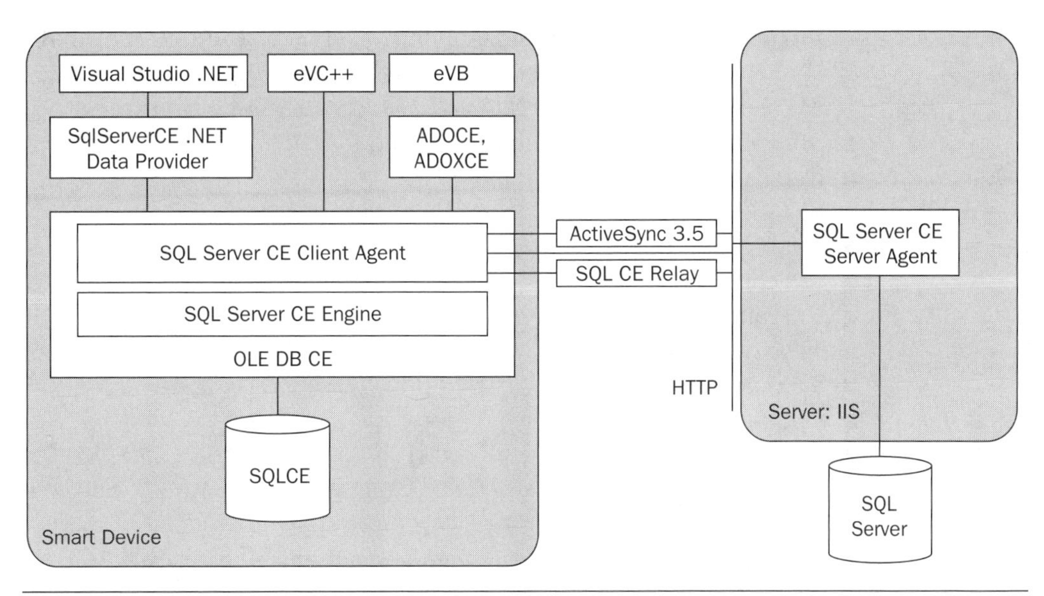

Figure 5–2 *SQLCE Architecture*. This diagram shows all the software components that make up an application that uses SQLCE. Note that both client and server components are required and that SQLCE can be accessed from managed and unmanaged code.

using eVC++. This provider encapsulates the SQL Server Client Agent that is responsible for replication and RDA, discussed further in Chapter 7, and the SQL Server CE Engine.

SQL Server CE Engine

KEY POINT

Unlike the server version, the SQLCE database engine is implemented in a DLL for performance reasons, even though the engine will be loaded in each process, using SQLCE on the device. However, because typically only one application using SQLCE will be active at any one time and because SQLCE supports only one concurrent connection, it is not a significant issue.

The database engine consists of two components: the storage engine that manages the data stored on the device in 4K pages and the query processor that processes (compiles, optimizes, and generates query plans) queries sent from applications. Together these two components support the following features, among others:

- Query processor supports SQL, including `SELECT`, `MAX`, `MIN`, `COUNT`, `SUM`, `AVG`, `INNER/OUTER JOIN`, `GROUP BY/HAVING`, `ORDER BY`, `UNION`,

and operators including ALL, AND, ANY, BETWEEN, EXISTS, NOT, SOME, OR, LIKE, IN, also Transact-SQL including DATEADD, DATEDIFF, GETDATE, COALESCE, SUBSTRING, and @@IDENTITY[5] among others
- 249 indexes per table, multicolumn indexes
- Databases of up to 2GB
- BLOBs of up 1GB
- 255 columns per table
- 128 character identifiers
- Unlimited nested subqueries
- Nested transactions
- Support for NULL values
- Parameterized queries
- Data Manipulation Language (DML): INSERT, UPDATE, and DELETE
- Data Definition Language (DDL): CREATE, DROP, ALTER on databases, tables, and indexes
- 17 data types, including Unicode (nchar, ntext) and GUID (uniqueidentifier)
- PRIMARY KEY, UNIQUE, and FOREIGN KEY constraints
- Replication tracking capability that tracks changed data on the device as discussed in Chapter 7

Query Analyzer

Although not shown in Figure 5-2, SQLCE also ships with an improved Query Analyzer that runs on the device and can be used to create databases, tables, and indexes; query data; insert and delete rows; compact and repair a database. It is generally used by developers to ensure that their local database is accessible.

When using the Compact Framework, Query Analyzer (Isqlw20.exe) is deployed to the device automatically using a .cab file if the application references the SqlServerCe provider, and a shortcut is placed in the Start menu on the device.[6]

To use the Query Analyzer, a developer need simply navigate to the database file (typically with an .sdf extension) and tap the green arrow on

---

[5] Both IDENTITY columns and uniqueidentifier can be used to create system-assigned primary key values in a SQLCE table. However, if all rows in the table are deleted and the database compacted, the identity counter is reset to its original value. This behavior can affect applications that need to synchronize with a back-end database.

[6] eVB and eVC++ developers must manually copy the Query Analyzer to the device, along with the appropriate supporting files, as noted in the documentation.

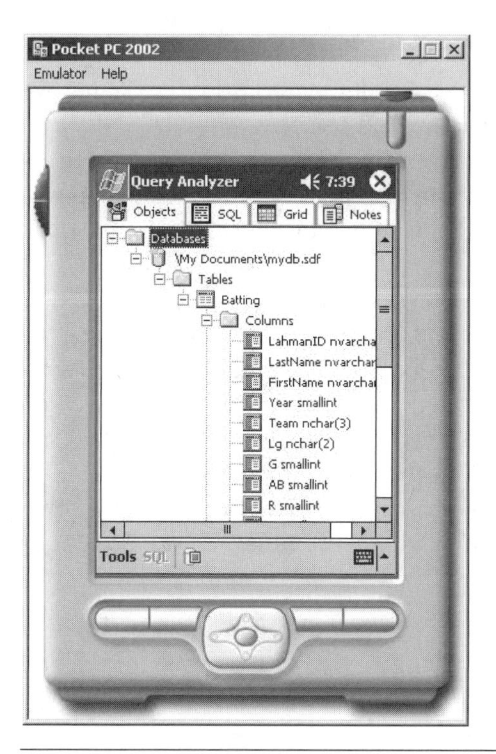

Figure 5–3 *Query Analyzer.* This screen shot shows navigating tables and columns in a local SQLCE database using the Query Analyzer.

the bottom of the screen. The tables and their structures can then be navigated, as shown in Figure 5-3. Developers can then query the data in a table by tapping on the arrow or tapping the SQL pane and writing the SQL directly.

Accessing SQLCE

To access SQLCE Compact Framework programmatically, developers can use the managed .NET Data Provider referred to as SqlServerCe. In this section we'll explore the data provider and how it can be used to connect to, query, and update SQLCE, and we'll also show a technique for writing provider-independent code when an application must access both a remote SQL Server and SQLCE.

SqlServerCe Provider Architecture

The SqlServerCe provider was implemented using the same pattern as the SqlClient .NET Data Provider used to access remote SQL Servers, as discussed in the previous chapter; therefore, it consists of the same basic classes. This parity with the desktop provider allows developers to leverage their existing ADO.NET knowledge and begin writing applications for SQLCE.

Unlike SqlClient, however, the SqlServerCe provider is shipped in an assembly separate from `System.Data.dll` and, so, must be explicitly referenced by the developer in his or her SDP. Figure 5-4 shows a diagram of the architecture of the provider, all of whose classes are found in the `System.Data.SqlServerCe` namespace.[7]

You'll notice in Figure 5-4 that the layout of the classes is similar to that found in Figure 4-4. For example, SqlServerCe supports both the disconnected programming model using the `DataSet` via the `SqlCeDataAdapter`

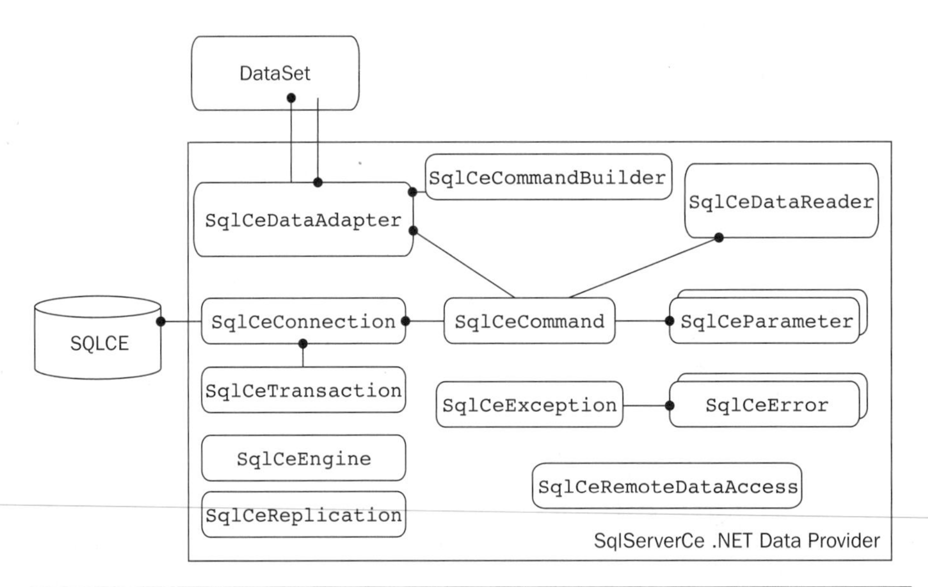

Figure 5–4 *SqlServerCe Architecture.* This diagram shows the primary classes found in the `System.Data.SqlServerCe` namespace in the SqlServerCe .NET data provider. Not shown are the `collection` and `events` classes, delegates, and enumerations.

[7] All the listings and code snippets in this chapter assume that the `System.Data.SqlServerCe` and `System.Data` namespaces have been imported (using C#).

object and the connected model using `SqlCeDataReader`. SQL commands are encapsulated with the `SqlCommand` class and can use parameters represented by `SqlCeParameter`. The `SqlCeConnection` and `SqlCeTransaction` objects also support local transactions, while database engine errors are captured in `SqlCeError` objects and thrown using a `SqlCeException` object. In fact, SqlServerCe even supports the `SqlCeCommandBuilder` class that can be used to create the `INSERT`, `UPDATE`, and `DELETE` statements automatically for synchronizing data in a data set with SQLCE. However, you'll also notice that SqlServerCe includes the additional classes shown in Table 5-1. These classes are found only in the SqlServerCe provider and have no analogs in SqlClient.

Although `SqlCeEngine` will be discussed in the following section, both `SqlCeReplication` and `SqlCeRemoteDataAccess` used for synchronization will be covered in detail in Chapter 7.

Manipulating Data with SqlServerCe

Once the SqlServerCe provider is referenced in an SDP, it can be used to manipulate SQLCE on the device. In this section we'll look at the common tasks developers will need to perform against SQLCE.

Creating Databases and Objects

Although a database with the appropriate structure and data can be deployed with the application, it is sometimes necessary for developers to create databases and objects on the fly. This can be accomplished using the `CreateDatabase` method of the `SqlCeEngine` object. In fact, a good strategy is to encapsulate the creation in a utility class and expose the functionality through shared methods like that shown in Listing 5-1.

Table 5–1 Additional SqlServerCe Classes

Namespace	Use
`SqlCeEngine`	Includes the methods and properties used to manipulate the SQL Server CE engine directly.
`SqlCeReplication`	Allows developers to use merge replication with SQL Server 2000; discussed fully in Chapter 7.
`SqlCeRemoteDataAccess`	Allows developers to access a data store remotely and synchronize its data with SQLCE; discussed fully in Chapter 7.

TIP: If you or your developers do elect to create a utility class to encapsulate common database functionality, you should consider marking the class as sealed (`NotInheritable` in VB) and giving it a private constructor. In this way, other developers can neither derive from the class nor create public instances of it. All the listings in this section can be thought of as methods in such a data-access utility class.

Listing 5–1 *Creating a SQLCE Database.* This method shows how to create a SQLCE database on the device using the `SqlCeEngine` class.

```
Public Shared Function CreateDb(ByVal filePath As String) As Boolean
   ' Delete and create the database

   Try
      If File.Exists(filePath) Then
          File.Delete(filePath)
      End If
   Catch e As Exception
      _lastException = e
      MsgBox("Could not delete the existing " & filePath, _
        MsgBoxStyle.Critical)
      Return False
   End Try

   Dim eng As SqlCeEngine
   Try
       eng = New SqlCeEngine("Data Source=" & filePath)
       eng.CreateDatabase()
       Return True
   Catch e As SqlCeException
       _lastException = e
       LogSqlError("CreateDb",e)
       MsgBox("Could not create the database at " & filePath, _
         MsgBoxStyle.Critical)
       Return False
   Finally
       eng.Dispose()
   End Try
End Function
```

In this case you'll notice that the `CreateDb` method first attempts to delete the database if it exists; it then passes the path to the database to the constructor of `SqlCeEngine` before calling the `CreateDatabase` method. The connection string need only consist of the `Data Source` attribute, and the `Provider` attribute will be defaulted to `Microsoft.SQLSERVER.OLEDB.CE.2.0`.[8] Other attributes may also be used, as discussed later in the chapter.

If an exception is found, the exception is placed in a private variable called `_lastException` that is exposed as a read-only shared property of the class. In this way the caller can optionally access full information about the exception that occurred. The database error is also logged using a custom method. To use this method the calling code would look like the following (assuming the method was placed in the `Atomic.SqlCeUtils` class):

```
If Atomic.SqlCeUtils.CreateDatabase(FileSystem.DocumentsFolder & _
   "\Personal\mydb.sdf") Then
    ' Go ahead and create some tables
End If
```

Note that the calling code uses the `FileSystem` class shown in Listing 3-5 to retrieve the My Documents folder on the device and then creates the database in the Personal folder.

NOTE: Databases may also be created using the `CREATE DATBASE` DDL statement when already connected to a different database. This statement also supports password protecting and encrypting the database, as discussed later in the chapter.

To create objects within a database, the application must first create a connection with the `SqlCeConnection` object. This is easily accomplished by passing the same connection string used to initialize the `SqlCeEngine` object in Listing 5-1 to the constructor of `SqlCeConnection` and calling the `Open` method as follows:

```
Dim cnCE As New SqlCeConnection(dbConnect)
cnCE.Open()
```

---

[8] This differs from ADOCE used in eVB, where omitting the `Provider` attribute assumes the `CEDB` provider and not `SQLCE`.

As you would expect, the previous snippet may throw a `SqlCeException` on either line if the connection string is malformed or the database is already open or does not exist. For this reason the opening of a connection should also be wrapped in a `Try-Catch` block.

More important, as SQLCE supports only one concurrent connection (unlike SQL Server 2000) because Windows CE is a single-user operating system, the connection object is usually obtained early in the run of the application and persisted in a variable until the application closes. It is therefore important to ensure that the connection eventually gets closed so that other applications (for example, the Query Analyzer) may connect to the database.

After creating a connection, DDL statements can be executed against the connection to create the appropriate tables and indexes. Each DDL statement must be encapsulated in a `SqlCeCommand` object and executed with the `ExecuteNonQuery` method. However, if the application requires that multiple statements be executed (to create several tables and their indexes, for example), it is possible to create a utility function to read the SQL from a resource file deployed with the application. This is accomplished by adding a text file to SDP and setting its Build Action property in the Properties window to Embedded Resource. Then the resource file can be populated with `CREATE` and `ALTER` statements, like those shown below, to create a table to hold batting statistics and add a primary key and an index.

```
CREATE TABLE Batting (Id int NOT NULL, LastName nvarchar(50),
 FirstName nvarchar(50),Year smallint NOT NULL,Team nchar(3),
 G smallint NULL,AB smallint NULL,R smallint NULL ,
 H smallint NULL,"2B" smallint NULL,"3B" smallint NULL ,
 HR smallint NULL,RBI smallint NULL);
ALTER TABLE Batting ADD CONSTRAINT pk_batting PRIMARY KEY (Id, Year);
CREATE INDEX idx_bat_team ON Batting (Year, Team ASC);
```

When the project is built, the file will then be compiled as a resource in the assembly and deployed to the device.

To read the resource script and execute its DDL, a method like that shown in Listing 5-2 can be written.

Listing 5–2 *Running a SQL Script*. This method reads from a resource file and executes all the commands found therein. Note that none of the commands may use parameters.

```
Public Shared Function RunScript(ByVal scriptName As String, _
  ByVal cn As SqlCeConnection) As Boolean
```

```
    ' Perform a simple execute non query
    Dim closeIt As Boolean = False
    Dim resource As Stream

    Try
        Resource = _
         [Assembly].GetExecutingAssembly().GetManifestResourceStream( _
         scriptName))
        Dim sr As New StreamReader(resource)
        Dim script As String = sr.ReadToEnd()
        Dim commands() As String
        commands = script.Split(";"c)

        ' Open the connection if closed
        If cn.State = ConnectionState.Closed Then
           cn.Open()
           closeIt = True
        End If

        Dim cm As New SqlCeCommand()
        cm.Connection = cn
        Dim s As String
        For Each s In commands
           If s <> "" Then
               cm.CommandText = s
               cm.ExecuteNonQuery()
           End If
        Next

    ' Clean up
    Catch e As SqlCeException
         _lastException = e
         LogSqlError("RunScript",e)
         MsgBox("Could not run script " & scriptName, _
          MsgBoxStyle.Critical)
        Return False
    Catch e As Exception
         _lastException = e
         MsgBox("Could not run script " & scriptName, _
          MsgBoxStyle.Critical)
        Return False
    Finally
       If closeIt Then cn.Close()
    End Try
    Return True
End Sub
```

In Listing 5-2 you'll notice that the GetManifestResourceStream method of the System.Reflection.Assembly class is used to read the resource file into a Stream object. The Stream object is then read by a StreamReader and placed into a string variable. In this scenario, the method is expecting strings delimited with a semicolon and, therefore, creates an array of strings using the Split method. This is required in order to execute multiple statements, because SQLCE does not support batch SQL as SQL Server does. In other words, SQLCE can execute only one statement per SqlCeCommand object.

The method then proceeds to open the connection object if it is closed and create a SqlCeCommand object. The command object is then populated repeatedly in a loop, and each statement is executed using ExecuteNonQuery. You'll notice in the Finally block that the connection is closed only if it were opened by the method. The advantage to this technique is that it allows for looser coupling between the script and the code that executes it, so that the script can be changed without changing any code and the project recompiled and deployed. To put it all together, an application could use code like the following in its main form's Load event to create the database, connect to it, and create tables and indexes:

```
If Atomic.SqlCeUtils.CreateDatabase(FileSystem.DocumentsFolder & _
   "\Personal\mydb.sdf") Then
    ' Connect (cnCE is global)
   cnCE = New SqlCeConnection(dbConnect)
   cnCE.Open()
   ' Go ahead and create some tables
   If Atomic.SqlCeUtils.RunScript("firstrun.sql", cnCE) Then
      ' All is well and the database is ready
   End If
End If
```

Querying Data

As mentioned previously, SqlServerCe supports both the disconnected and connected programming models using the DataSet and data reader that were discussed in Chapter 4. Unfortunately, unlike in SQL Server 2000, SQLCE does not support stored procedures. As a result, developers will need to formulate SQL within the application and submit it to the database engine (although SQLCE does support parameterized queries, as will be discussed later). Also, as mentioned previously, SQLCE does not support batch SQL, and so, multiple SELECT statements cannot be executed and their results cannot be automatically populated in multiple DataTable objects in a DataSet or through multiple result sets using the NextResult property of the SqlCeDataReader. However, developers can still create data-access helper methods that reduce the amount of code required by the caller. For example, the method in Listing 5-3 adds data to a DataSet based on the SQL passed to the method.

Listing 5–3 *Populating a Data Set.* This method adds data to a data set given the SQL statement and the connection object to use.

```
Public Shared Sub FillSimpleDataSet(ByVal ds As DataSet, _
   ByVal sql As String, ByVal cn As SqlCeConnection, _
   ByVal acceptChanges As Boolean)

   Try
       Dim cm As New SqlCeCommand(sql, cn)
       Dim da As New SqlCeDataAdapter(cm)
       da.AcceptChangesDuringFill = acceptChanges

       da.MissingMappingAction = MissingMappingAction.Passthrough
       da.MissingSchemaAction = MissingSchemaAction.AddWithKey

       da.Fill(ds)
   Catch e As SqlCeException
       LogSqlError("FillSimpleDataSet",e)
       Throw New SqlCEUtilException( _
         "Could not fill dataset for: " & sql, e)
   End Try
End Function
```

You'll notice that in Listing 5-3 an existing connection object is used and that the caller determines whether AcceptChangesDuringFill is set to True or False to determine if the newly added rows are treated as new rows (with their RowState property set to Added) or as unmodified rows. In this case the connection object needn't be opened explicitly because the SqlCe-DataAdapter will open it if it is not already open. The MissingMapping-Action and MissingSchemaAction properties are also set to allow the data adapter to create any missing tables or columns in the DataSet and to add primary key information if available. Obviously, this method would not be useful if more sophisticated table mappings were required.[9] If any errors

[9] See Chapter 12 of *Teach Yourself ADO.NET in 21 Days,* by Dan Fox, for a complete explanation of how data adapters use table and column mappings.

occur, a custom exception of type SqlCeUtilException inherited from ApplicationException is thrown.

NOTE: Creating custom exception classes like SqlCeUtilException in Listing 5-3 that can be used to encapsulate application-specific messages and custom methods and properties is a good strategy. The original exception can then be chained to the custom exception using the InnerException property. This technique of exception wrapping, or chaining, allows the application to add specific messages at multiple levels in the call stack.

Data readers can similarly be created to stream through the results from a table as shown in Listing 5-4.

Listing 5–4 *Creating a Data Reader.* This method creates and returns a SqlCeDataReader given a SQL statement and a connection object.

```
Public Shared Function ExecDataReader(ByVal sql As String, _
   ByVal cn As SqlCeConnection) As SqlCeDataReader

    Try
         ' Create the command
         Dim cm As New SqlCeCommand(sql, cn)
         If cn.State = ConnectionState.Closed Then
             cn.Open()
         End If

         ' Execute data reader
         Dim dr As SqlCeDataReader
         dr = cm.ExecuteReader()
         Return dr
    Catch e As SqlCeException
         LogSqlError("ExecDataReader",e)
         Throw New SqlCEUtilException( _
           "Could not execute data reader for :" & sql, e)
    End Try
End Function
```

In Listing 5-4 the method creates a command object and associates it with the connection passed into the method. In this case the method must also open the connection if it is not already open before executing the data reader and returning it. Note that although the `ExecuteReader` method supports the `CloseConnection` and other command behaviors, it is not used because typically a single global database connection remains open for the lifetime of the application.

A caller would then use the method as follows:

```
Dim dr As SqlCeDataReader
dr = SqlCeUtils.ExecDataReader( _
  "SELECT * FROM Batting WHERE Id = 660", cnCE)

Do While dr.Read()
   ' Process the data
Loop
dr.Close()
```

Although not shown in this listing, it is also interesting to note that unlike the SqlClient provider, the SqlServerCe provider does support multiple data readers on the same open connection object. In other words, developers needn't close the `SqlCeDataReader` before using the connection to execute another command. Again, this is the case because SQLCE supports only a single concurrent connection.

One of the most interesting new features of SQLCE is the inclusion of parameterized queries. Using parameterized queries, developers can simply populate `SqlCeParameter` objects associated with a `SqlCeCommand`, rather than having to manually concatenate parameters into a single string. In addition, parameterized queries are recommended for performance reasons. However, unlike SqlClient, SQLCE supports only positional parameters, and the parameters must be defined in the SQL statement using a question mark. In other words, developers must declare a `SqlCeParameter` object for each question mark in the SQL statement so that the `SqlCeCommand` object can perform the substitution at runtime. For example, in order to execute the query shown above as a parameterized query, a developer could do the following:

```
Dim dr As SqlCeDataReader
Dim cm As New SqlCeCommand("SELECT * FROM Batting WHERE Id = ?", cnCE)
cm.Parameters.Add(New SqlCeParameter("@Id", SqlDbType.Int))
cm.Parameters(0).Value = 660

dr = SqlCeUtils.ExecDataReader()
```

In this case, although the parameter was referenced by its ordinal, it could alternatively have been referenced by its name (`@Id`).

In a helper or utility class, the creation of parameter objects and their association with a command object can be handled by a structure and private method like that shown in Listing 5-5.

Listing 5–5 Listing 5-5: *Automating Parameterized Queries.* This structure and method can be used to create and attach parameters automatically to a `SqlCeCommand` object.

```
Public Structure ParmData
    Public Name As String
    Public Value As Object
    Public DataType As SqlDbType

    Public Sub New(ByVal name As String, ByVal dataType As SqlDbType, _
      ByVal value As Object)
        Me.Name = name
        Me.DataType = dataType
        Me.Value = value
    End Sub
End Structure

Private Shared Function PopulateCommand(ByVal sql As String, _
   ByVal parms As ArrayList, ByVal cn As SqlCeConnection) _
   As SqlCeCommand

    Dim cm As New SqlCeCommand(sql, cn)
    cm.CommandType = CommandType.Text

    ' Populate parameters
    Dim p As Object
    For Each p In parms
        Dim p1 As ParmData = CType(p, ParmData)
        cm.Parameters.Add( _
          New SqlCeParameter(p1.Name, p1.DataType, p1.Value))
    Next

    Return cm

End Function
```

In Listing 5-5 you'll notice that the private **PopulateCommand** method accepts an **ArrayList** of **ParmData** objects as a parameter and uses it to populate a **SqlCeCommand** created from the SQL statements and **SqlCeConnection** object passed in as well.[10] With this technique an overloaded version of the method in Listing 5-4 can be created to accept parameterized SQL, as shown in Listing 5-6.

Listing 5–6 *Creating a Data Reader with Parameters.* This method creates and returns a SqlCeDataReader given a SQL statement, parameters, and a connection object.

```
Public Shared Function ExecDataReader(ByVal sql As String, _
      ByVal cn As SqlCeConnection, _
      ByVal parms As ArrayList) As SqlCeDataReader

    Try
         ' Create the command
         Dim cm As SqlCeCommand = Me.PopulateCommand(sql, parms, cn)
         If cn.State = ConnectionState.Closed Then
             cn.Open()
         End If

         ' Execute data reader
         Dim dr As SqlCeDataReader
         dr = cm.ExecuteReader()
         Return dr
    Catch e As SqlCeException
         LogSqlError("ExecDataReader",e)
         Throw New SqlCEUtilException( _
           "Could not execute data reader for :" & sql, e)
    End Try
End Function
```

At this point the caller need create only the **ParmData** objects, specifying the appropriate data type, and place them in an **ArrayList** before passing them to **ExecDataReader**, as shown in this snippet:

[10] Since SQLCE does not support stored procedures, the **StoredProcedure CommandType** is also not supported.

```
Dim dr As SqlCeDataReader
Dim sql As String = "SELECT * FROM Batting WHERE Id = ?"
Dim parms As New ArrayList()

parms.Add(New ParmData("id", SqlDbType.Int, 660))
dr = SqlCeUtils.ExecDataReader(sql, cnCE, parms)
```

Using Indexes

Perhaps the biggest difference between the SqlClient provider and the SqlServerCe provider is the inclusion of index seeks using data readers in SqlServerCe. Using this technique allows developers to write code that performs better than issuing `SELECT` statements with `WHERE` clauses. This is the case because the SQLCE query processor must compile, optimize, and generate a query plan for each query, while performing the index seek directly avoids these costly steps. The caveat is that this works only against single tables, and the table must of course have an index. As a result, for complex queries developers will likely want to rely on the query processor.

NOTE: In one example documented on Microsoft's SQLCE Web site and referenced in the "Related Reading" section, using an index seek versus the query processor improved performance by a factor of 20 or greater.

For example, consider the scenario where a developer wanted to retrieve the statistics for a specific team and year from the batting table created in Listing 5-2, and it is known that the year will be in the range from 1980 to 1989. The batting table has a composite index on the Year and Team columns, and so a method like that shown in Listing 5-7 can be written to return a `SqlCeDataReader` positioned on the correct row.

Listing 5–7 *Seeking a Row Using an Index.* This method creates and returns a `SqlCeDataReader` positioned on the appropriate row for a given set of index values.

```
public static SqlCeDataReader ExecTeamReader(SqlCeConnection cn,
  string team, int year)
{
   SqlCeCommand cmd = new SqlCeCommand("Batting",cn);
   cmd.CommandType  = CommandType.TableDirect;
```

```
    if (cn.State == ConnectionState.Closed)
    {
       cn.Open();
    }

    // Index contains Year and Team
    cmd.IndexName = "idx_bat_team";

    object[] start = {1980, 1989};
    object[] end = {null, null};
    cmd.SetRange(DbRangeOptions.InclusiveStart |
      DbRangeOptions.InclusiveEnd, start, end);

    Try
    {
       SqlCeDataReader rdr = cmd.ExecuteReader();
       rdr.Seek(DbSeekOptions.AfterEqual, year, team);
       return rdr;
    }
    Catch (SqlCeException e)
    {
      LogSqlError("ExecTeamReader",e);
      // Throw a custom exception
      return null;
    }
}
```

You'll notice in Listing 5-7 that the `SqlCeCommand` must have its `CommandText` property set to the name of the table to search and that the `CommandType` must be set to `TableDirect`. The name of the index is then set using the `IndexName` property. Although it is not required, this listing also shows that the range of values searched can be restricted by passing arrays of start and end values to the `SetRange` method. The `DbRangeOptions` enumeration determines how the `Seek` method uses the start and end values. After opening the data reader using `ExecuteReader`, its `Seek` method is then called with a value from the `DbSeekOptions` enumeration. This value specifies which row if any is to be returned. In this case, `AfterEqual` is used and if a row is not found, the first row after the index range will be the one pointed to by the data reader. Alternatively, if `FirstEqual` is used, the `Seek` method will throw a `SqlCeException` if a row cannot be located.

A caller can then invoke the method to position a data reader at the statistics for the 1984 Chicago Cubs as follows:

```
SqlCeDataReader dr = SqlCeUtils.ExecTeamReader(cnCE,"CHN",1984);
```

Modifying Data

Inserting, updating, and deleting data in SQLCE are not handled any differently than they are using the SqlClient provider, with the exception, of course, that SQLCE does not support stored procedures. In other words developers may use the `SqlCeDataAdapter` to modify data in an underlying base table utilizing the table and column mappings collections and then invoking the `Update` method of the data adapter. Developers may also execute command objects directly. In either case parameterized queries are used and, in fact, are required for use with the `SqlCeDataAdapter`.

For example, to insert a new row into the Batting Table, the method shown in Listing 5-8 could be written to return the command object used in either scenario.

Listing 5–8 *Inserting Data with a Command.* This method creates and returns a `SqlCeCommand` to insert new rows into the Batting Table.

```
Public Shared Function GetBattingCmd(cnCE As SqlCeConnection, _
   trans As SqlCeTransaction) As SqlCeCommand

   Dim sql As String = "INSERT INTO Batting (Id, LastName, " & _
     "FirstName, Year, Team, G, AB, R, H, ""2B"", ""3B"", " & _
     "HR, RBI) VALUES (?, ?, ?, ?, ?, ?, ?, ?, ?, ?, ?, ?, ?)"

   battingCmd = New SqlCeCommand(sql)
   battingCmd.CommandType = CommandType.Text
   If Not trans Is Nothing Then
       battingCmd.Transaction = trans
   End If

   battingCmd.Parameters.Add(New SqlCeParameter("Id", _
     SqlDbType.NVarChar, 9, "Id"))
   battingCmd.Parameters.Add(New SqlCeParameter("LastName", _
     SqlDbType.NVarChar, 50, "LastName"))
   battingCmd.Parameters.Add(New SqlCeParameter("FirstName", _
     SqlDbType.NVarChar, 50, "FirstName"))
   battingCmd.Parameters.Add(New SqlCeParameter("Year", _
     SqlDbType.SmallInt, 4, "Year"))
```

```
    battingCmd.Parameters.Add(New SqlCeParameter("Team", _
      SqlDbType.NVarChar, 3, "Team"))
    battingCmd.Parameters.Add(New SqlCeParameter("G", _
      SqlDbType.SmallInt, 4, "G"))
    battingCmd.Parameters.Add(New SqlCeParameter("AB", _
      SqlDbType.SmallInt, 4, "AB"))
    battingCmd.Parameters.Add(New SqlCeParameter("R", _
      SqlDbType.SmallInt, 4, "R"))
    battingCmd.Parameters.Add(New SqlCeParameter("H", _
      SqlDbType.SmallInt, 4, "H"))
    battingCmd.Parameters.Add(New SqlCeParameter("2B", _
      SqlDbType.SmallInt, 4, "2B"))
    battingCmd.Parameters.Add(New SqlCeParameter("3B", _
      SqlDbType.SmallInt, 4, "3B"))
    battingCmd.Parameters.Add(New SqlCeParameter("HR", _
      SqlDbType.SmallInt, 4, "HR"))
    battingCmd.Parameters.Add(New SqlCeParameter("RBI", _
     SqlDbType.SmallInt, 4, "RBI"))

    return battingCmd
End Function
```

The GetBattingCmd static method could then be used by a caller to retrieve the appropriate command before populating the parameters with values manually through code or by setting it to the InsertCommand property of the SqlCeDataAdapter.

Although not typically recommended for production scenarios with the SqlClient provider,[11] the command builder included with SqlServerCe (SqlCeCommandBuilder) can be used in place of code like that shown in Listing 5-7. This is due to the fact that SQLCE is single user and runs in process with the application; therefore, an extra round trip isn't as costly in terms of performance. In any case, it can be used simply by passing the SqlCeDataAdapter to the constructor of the command builder:

```
Dim cb as New SqlCeCommandBuilder(da)
```

When needed,[12] the command builder will then build the insert, update, and delete commands based on the SELECT statement exposed in

[11] Using the SqlCommandBuilder object always engenders one extra trip to the database server so that the command builder can determine column names and data types.

[12] When the RowUpdating events fire on the SqlCeDataAdapter object.

the CommandText property of the SelectCommand. Note that just as with the SqlClient provider, the SELECT statement used by the data adapter mustn't be complex (contain aggregates columns and joins) and must return at least one primary key or unique column, or an exception will be thrown.

NOTE: If the SELECT statement changes or the connection or transaction associated with the command changes, a developer can call the RefreshSchema method of the SqlCeCommandBuilder to regenerate the insert, update, and delete commands.

Handling Transactions

Just like SqlClient, the SqlServerCe provider supports transactions, or the ability to group a series of data modifications in an atomic operation. This is useful if an application needs to update two tables, with the requirement that if one of the updates fails, they both fail (a parent/child relationship, for example).

This is accomplished through the BeginTransaction method of the SqlCeConnection object, which spawns a SqlCeTransaction object used to control the outcome (Commit or Rollback) of the transaction. For example, the following code snippet uses the GetBattingCmd method shown in Listing 5-8 and the GetPitchingCmd method (not shown) to execute two commands in a single transaction:

```
SqlCeTransaction trans = null;
Try
{
    trans = cnCE.BeginTransaction();
    SqlCeCommand bat = SqlCeUtils.GetBattingCmd(cnCE, trans);
    SqlCeCommand pitch = SqlCeUtils.GetPitchingCmd(cnCE, trans);

    // populate the commands with the new values

    bat.ExecuteNonQuery();
    pitch.ExecuteNonQuery();
    trans.Commit();
}
catch (SqlCeException e)
{
  if (trans != null) {trans.Rollback();}
  LogSqlError("MyMethod",e);
  // most likely throw a custom exception
}
```

You'll notice here that the `GetBattingCmd` and `GetPitchingCmd` methods accept a transaction as the second argument. Referring to Listing 5-8, this transaction, if instantiated, is associated with the command object using its `Transaction` property.

However, the transactional behavior of SQLCE differs from SQL Server 2000, and so developers must be aware of four differences. First, SQLCE only supports an isolation level of `ReadCommitted`, which exclusively locks data being modified in a transaction. As a result, the `IsolationLevel` property of `SqlCeTransaction` can only be set to the `ReadCommitted` value of the `IsolationLevel` enumeration. Second, SQLCE supports nested transactions, but only up to five levels. Third, SQLCE holds an exclusive lock on any table that has been modified in a transaction.[13] This means that any attempt to access any data from the table outside the transaction, while it is pending, will result in an exception. Fourth, if a data reader is opened within a transaction, the data reader will automatically be closed if the transaction is rolled back. If the transaction commits, the data reader can still be used.

Abstracting .NET Data Providers

As discussed in Chapter 4, applications written with VS .NET 2003 and the Compact Framework can access a SQL Server 2000 server remotely using the SqlClient .NET Data Provider. And, as discussed in this chapter, applications can store data locally in SQLCE using the SqlServerCe .NET Data Provider. In some scenarios, an application may wish to do both, for example, by accessing the remote SQL Server when connected to a corporate LAN via a direct connection, WLAN, or WAN and accessing SQLCE when disconnected.

In these instances, developers can take advantage of the object-oriented nature of the Compact Framework to write code that can be used with either provider. Doing so allows a greater level of code reuse and easier porting of code from the desktop Framework to the Compact Framework. Abstracting data providers is possible since, as mentioned in Chapter 4, all .NET Data Providers are implemented using the same underlying base classes and interfaces. These include the interfaces `IDbConnection`, `IDbCommand`, `IDataRecord`, `IDataParameter`, `IDbDataParameter`, `IDataParameter-Collection`, `IDataReader`, `IDataAdapter`, `IDbDataAdapter`, and `IDb-Transaction`, along with the `DataAdapter` and `DbDataAdapter` classes,

[13] As opposed to row- and page-level locks used by SQL Server 2000.

Interfaces or Base Classes?

The Compact Framework relies on both interfaces and base classes to allow code reuse through inheritance and polymorphism. Simply put, interfaces (typically prefixed with an "I") enable interface inheritance by allowing a class to implement a set of method signatures defined in the interface. When using interface inheritance, the class implementing the interface must include all of the method signatures from the interface but must implement the functionality of the methods itself. Using a base class, a class may use implementation inheritance to inherit both the method signatures and the implementation (the code) in the base class. The derived class may then override the methods of the base class to augment or replace the base class code.

Both techniques are useful, and, as you would imagine, interface inheritance is used when a variety of different classes needs to implement the same behavior (methods) in different ways, while implementation inheritance is used when classes form a natural hierarchy represented with an "is a" relationship (Employee is a Person). Both can be used together in the same class, although in the Compact Framework, implementation inheritance is restricted to a single inheritance, meaning that each class may inherit only from one base class.

Using both techniques, developers can write polymorphic (literally "multiform") code by targeting the reference variables in their code at the interfaces and base classes, rather than at the class inheriting from the interface or base class (often called the concrete class). In this way, at runtime the reference variables may actually refer to instances of any of the concrete classes in the inheritance relationship, thereby allowing the code to work in a variety of scenarios.

among others, found in the `System.Data` and `System.Data.Common` namespaces.

One technique for abstracting the data provider used is to implement the Abstract Factory design pattern documented in the book *Design Patterns*, as noted in the "Related Reading" section at the end of the chapter. This design pattern allows code to create families of related classes without specifying their concrete classes at design time. In this case, the family of related classes comprises the classes that make up a data provider, including connection, command, data adapter, and parameter.

Although it is possible to use the Abstract Factory pattern as documented in *Design Patterns*, a slight variant of the pattern, shown in Listing 5-9, is flexible because it allows the data provider to be specified in a shared method of the Abstract Factory class rather than having to be hard-coded at the creation of the class at runtime.

Listing 5–9 *Implementing the Abstract Factory Pattern.* This listing shows the code necessary to implement the Abstract Factory pattern so that polymorphic code can be written to use either of the data providers that ships with the Compact Framework. Note that the `SqlClientFactory` class is not shown.

```
Public Enum ProviderType
    SqlClient = 0
    SqlServerCe = 1
End Enum

Public MustInherit Class ProviderFactory

    Public Shared Function CreateFactory( _
      ByVal provider As ProviderType) As ProviderFactory
        If provider = ProviderType.SqlClient Then
            Return New SqlClientFactory
        Else
            Return New SqlServerCeFactory
        End If
    End Function

    Public MustOverride Function CreateConnection( _
      ByVal connect As String) As IDbConnection
    Public MustOverride Overloads Function CreateDataAdapter( _
      ByVal cmdText As String, _
      ByVal connection As IDbConnection) As IDataAdapter
    Public MustOverride Overloads Function CreateDataAdapter( _
      ByVal command As IDbCommand) As IDataAdapter
    Public MustOverride Overloads Function CreateParameter( _
      ByVal paramName As String, _
      ByVal paramType As DbType) As IDataParameter
    Public MustOverride Overloads Function CreateParameter( _
      ByVal paramName As String, _
      ByVal paramType As DbType, _
      ByVal value As Object) As IDataParameter
    Public MustOverride Function CreateCommand( _
       ByVal cmdText As String, _
       ByVal connection As IDbConnection) As IDbCommand

End Class

Public NotInheritable Class SqlServerCeFactory
   Inherits ProviderFactory
```

```
    Public Overrides Function CreateConnection( _
     ByVal connect As String) As IDbConnection
        Return New SqlCeConnection(connect)
    End Function

    Public Overloads Overrides Function CreateDataAdapter( _
      ByVal cmdText As String, _
      ByVal connection As IDbConnection) As IDataAdapter
        Return New SqlCeDataAdapter(cmdText, _
         CType(connection, SqlCeConnection))
    End Function

    Public Overloads Overrides Function CreateDataAdapter( _
      ByVal command As IDbCommand) As IDataAdapter
        Return New SqlCeDataAdapter(CType(command, SqlCeCommand))
    End Function

    Public Overloads Overrides Function CreateParameter( _
      ByVal paramName As String, _
      ByVal paramType As DbType) As IDataParameter
        Return New SqlCeParameter(paramName, paramType)
    End Function

    Public Overloads Overrides Function CreateParameter( _
      ByVal paramName As String, _
      ByVal paramType As DbType, _
      ByVal value As Object) As IDataParameter
        Dim parm As New SqlCeParameter(paramName, paramType)
        parm.Value = value
        Return parm
    End Function

    Public Overrides Function CreateCommand(ByVal cmdText As String, _
      ByVal connection As IDbConnection) As IDbCommand
        Return New SqlCeCommand(cmdText, _
         CType(connection, SqlCeConnection))
    End Function

End Class
```

As you'll notice in Listing 5-9, the `ProviderType` enumeration identifies which factory classes are available. The heart of the listing is the abstract (marked as **`MustInherit`** in VB and **`abstract`** in C#) **`ProviderFactory`** class. This class implements a shared method to create an instance of a concrete **`ProviderFactory`** class, along with a set of method signatures marked with the **`MustOverride`** keyword. This keyword ensures that the class inheriting from **`ProviderFactory`** will override the methods to provide an implementation. The **`SqlServerCeFactory`** class inherits from **`ProviderFactory`**, overriding the base class methods and returning instances of the appropriate SqlServerCe objects (**`SqlCeConnection`**, **`SqlCeDataAdapter`**, and so forth). Note that the methods of the **`ProviderFactory`** class return references to the interfaces implemented by data providers discussed earlier. This is the key to enabling the writing of polymorphic code. Although not shown in the listing due to space constraints, there would, of course, be a corresponding factory class for the SqlClient provider that also inherits from **`ProviderFactory`**.

NOTE: To extend the `ProviderFactory` to support new providers (for example, one for Sybase SQL Anywhere Studio), a developer need only create a factory class that inherits from `ProviderFactory`. He or she would also likely want to extend the `ProviderType` enumeration and the `CreateFactory` method.

To use the **`ProviderFactory`** class, a caller need only instantiate the correct class using the shared method, as follows:

```
Dim pf As ProviderFactory
If CheckForNetworkConn() Then
    ' Go remote
    pf = ProviderFactory.CreateFactory(ProviderType.SqlClient)
Else
    ' Go local
    pf = ProviderFactory.CreateFactory(ProviderType.SqlServerCe)
End If
```

In this snippet the **`CheckForNetworkConn`** method shown in Chapter 4 is used first to determine if a network connection is available; if so, it uses SqlClient and if not, SqlServerCe. Of course, the value for the **`ProviderType`** enumeration could also easily be read from a configuration file or passed into the method as a variable to allow for flexibility.

Once the concrete **`ProviderFactory`** has been created, it can be passed into methods like those shown in the listings in this chapter so that the methods can be used against either provider. For example, the `ExecDataReader` method shown in Listing 5-4 could then be rewritten as shown in Listing 5-10.

Listing 5–10 *Using the Abstract Factory Pattern.* This method shows the `ExecDataReader` method rewritten to use an instance of the `ProviderFactory` class to enable provider-independent database access.

```
Public Shared Function ExecDataReader(ByVal pf As ProviderFactory, _
   ByVal sql As String, ByVal cn As IDbConnection) As IDataReader

    Try
        ' Create the command
        Dim cm As IDbCommand = pf.CreateCommand(sql, cn)
        If cn.State = ConnectionState.Closed Then
            cn.Open()
        End If

        ' Execute data reader
        Dim dr As IDataReader
        dr = cm.ExecuteReader()
        Return dr
    Catch e As Exception
        LogSqlError("ExecDataReader",e)
        Throw New Exception( _
          "Could not execute data reader for :" & sql, e)
    End Try
End Function
```

Note that because the `ExecDataReader` method can now be used with either provider, it returns an object that implements the `IDataReader` interface and accepts an `IDbConnection` object, rather than the concrete types for SqlServerCe. In addition, the creation of the `SqlCeCommand` object has been replaced with a call to the `CreateCommand` method of the `ProviderFactory`, and the reference to the `SqlCeException` object in the `Catch` block has been replaced with the generic `Exception` object.[14]

[14] An alternative and more dynamic approach to creating an abstract factory class using the runtime type creation methods of the desktop Framework and Compact Framework can be found in Chapter 18 of *Teach Yourself ADO.NET in 21 Days*.

Administering SQLCE

Because SQLCE is separate from the Compact Framework, it must be administered separately. The administration tasks take the form of security administration, database maintenance, and installation and deployment.

Security

KEY POINT

As with any database, it is important that the data in SQLCE be secure. This is particularly the case because the device on which SQLCE is running is inherently mobile and can easily fall into the hands of someone who is not the intended user. As a result, it is important that Compact Framework applications be able to present an authentication dialog to users before providing access to the data and that the data itself can be encrypted on the device.

NOTE: Keep in mind that because Windows CE is a single-user operating system, there is no support in SQLCE for individual user authentication or permissions; and, in fact, the `syslogins`, `sysprotects`, and `sysusers` system tables present in SQL Server 2000 to support these functions are not included in SQLCE. Any user who can open the database has full permissions. Along the same lines, the Windows CE file system does not support permissions; so, there is no inherent protection for the .sdf file.

SQLCE supports these requirements by offering both password protection for the entire database file and encryption for the entire file using a 128-bit key.

Password Protection

Password protecting a SQLCE database can be done only when the database is created or compacted (as discussed in the next section) and can be done with either the `CreateDatabase` method of the `SqlCeEngine` object or the `CREATE DATABASE` DDL statement.

When using the `CreateDatabase` method, the password attribute is simply appended to the connection string passed into the constructor of the `SqlCeEngine` class. As a result, the `CreateDb` method shown in Listing 5-1 could be altered as shown in the following snippet to accept a password of up to 40 characters to use when creating the database.

```
Public Shared Function CreateDb(ByVal filePath As String, _
  ByVal pwd As String) As Boolean

   ' Code ommitted for brevity

   Dim eng As SqlCeEngine
   Try
       eng = New SqlCeEngine("Data Source=" & filePath & _
        ";password= & pwd)
       eng.CreateDatabase()
       Return True
   Catch e As SqlCeException
       ' Code ommitted for brevity
   End Try
End Function
```

Once the password has been created, there is no way to recover it; however, the password can be changed by compacting the database, as will be discussed later in this section.

If the application is executing DDL to create a database, a `CREATE DATABASE` statement like the following can be issued:

```
CREATE DATABASE 'mydb.sdf' DATABASEPASSWORD 'sdfg53$h'
```

Encryption

Just as with password protection, encrypting a SQLCE database can be accomplished with the `CreateDatabase` method, the process of compacting, or the `CREATE DATABASE` DDL statement.

To encrypt using `CreateDatabase`, the `encrypt database` attribute needs to be added to the connection string in addition to the password, as shown in the following snippet, where the `CreateDb` method from Listing 5-1 is once again modified to support an argument to determine if the database should be encrypted. Note, however, that the attribute needn't be provided when the database is opened.

```
Public Shared Function CreateDb(ByVal filePath As String, _
  ByVal pwd As String, ByVal encrypt As Boolean) As Boolean

   ' Code ommitted for brevity

   Dim eng As SqlCeEngine
```

```
    Try
        Dim connect = "Data Source=" & filePath & _
         ";password= & pwd
        If encrypt Then
            connect &= ";encrypt database=TRUE"
        End If
        eng = New SqlCeEngine(connect)
        eng.CreateDatabase()
        Return True
    Catch e As SqlCeException
        ' Code ommitted for brevity
    End Try
End Function
```

KEY POINT

The password attribute must be included because SQLCE uses the MD5[15] hashing algorithm to create the 128-bit key required by the RC4[16] algorithm used to encrypt the database. For this reason it is important that the password chosen be of a reasonable length to avoid easy cracking by hackers.[17] Although it would be cumbersome to force users to input 40-character passwords, passwords of at least 8 characters (including letters, numbers, and at least once special character) should suffice to offer a reasonable amount of protection. Changing passwords periodically via compaction is also a good strategy because it moves the target for any potential hacker.

To encrypt the database file using the `CREATE DATABASE` statement, the `ENCRYPTION ON` clause is used as follows:

```
CREATE DATABASE 'mydb.sdf' DATABASEPASSWORD 'sdfg53$h' ENCRYPTION ON
```

Database Maintenance

KEY POINT

As alluded to earlier, the `SqlCeEngine` class also supports the `CompactDatabase` method, which can be used to compact and reclaim wasted space that collects in the database as data and objects are deleted and tables are

---

[15] A message-digest algorithm developed in 1991 by RSA Security.

[16] A symmetric encryption algorithm designed by RSA Security in 1987 and used in Secure Sockets Layer (SSL) and other commercial applications.

[17] Hackers can extract the hash value from the .sdf file and then run either a dictionary or a brute-force attack to discover the password. Longer passwords are recommended because the effort required in using a brute-force method increases exponentially. For example, two-character passwords take seconds to break, while eight-character passwords can require years.

reindexed. It is recommended that SQLCE databases be periodically compacted because this also leads to improved query performance through index reordering and the refreshing of statistics used by the query processor to generate execution plans.

Compacting a database can also be used to change the collating order,[18] encryption, or password for the database, as mentioned previously in this section. This method creates a new database and requires that the source database be closed and that the destination file not exist. It is also important to remember that because a copy is created, the device will need to have enough room to make the copy or an error will result.

Once again, it makes sense to wrap the `CompactDatabase` functionality in a method that checks for the existence of the source database and then automatically copies the destination back to the source when completed, as shown in Listing 5-11, which takes advantage of the `FileSystem` class in Listing 3-5 to create the temporary destination that is ultimately moved back to the original file name.

Listing 5–11 *Compacting a SQLCE Database.* This method compacts a database, reclaiming wasted space, and copies the newly created database back to the old name.

```
Public Shared Function CompactDb(ByVal filePath As String) As Boolean

   If Not File.Exists(filePath) Then
      MsgBox("Source database does not exist = " & filePath, _
        MsgBoxStyle.Critical)
      Return False
   End If

   Dim eng As SqlCeEngine
   Try
       eng = New SqlCeEngine("Data Source=" & filePath)
       eng.Compact("Data Source=" & _
        FileSystem.GetSpecialFolderPath(ceFolders.PERSONAL) & _
        "\temp000.sdf")
       File.Delete(filePath)
```

[18] If not specified in the `CREATE DATABASE` statement or the destination database connection string, the default collation assigned is `Latin1_General`. This collation uses Latin 1 General dictionary sorting rules, code page 1,252, and is case-insensitive and accent-insensitive. All databases in SQLCE are always case-sensitive and accent-insensitive. To see the available collations, see the Books Online for SQLCE.

```
        File.Move(FileSystem.GetSpecialFolderPath( _
          ceFolders.PERSONAL) & "\temp000.sdf", filePath)
    Catch e As Exception
        _lastException = e
        MsgBox("Could not compact the database at " & filePath, _
         MsgBoxStyle.Critical)
        Return False
    Finally
        eng.Dispose()
    End Try

    Return True

End Function
```

It should also be noted that SQLCE creates a temporary file each time the database engine is initialized and attempts to delete it when the engine terminates normally. This file is used for storing pages that exceed the SQLCE buffer cache, as well as interim results and tables used in queries. By default, the file is created in the Temp directory on the device, although its location can be specified using the `temp file directory` attribute of the connection string as shown here:

```
Dim connect = "Data Source=\mydb.sdf;temp file directory=\StorageCard"
Dim eng As New SqlCeEngine(connect)
```

This may be required if the need to store the temporary file on a storage card, rather than in RAM, arises. The file will grow the most when transactions and large `UPDATE` and `DELETE` statements are executed. However, keep in mind that accessing storage cards is typically slower than accessing RAM; so, query performance may suffer as a result.

Installation and Deployment

To use SQLCE in a solution, components must be installed both on the development machines as well on the device. Fortunately for Compact Framework developers, all the required SQLCE components are installed and configured with VS .NET 2003. This allows a developer to reference the `System.Data.SqlServerCe.dll` assembly from any SDP and begin coding against SQLCE.

When an SDP that accesses SQLCE is deployed to either an emulator or an actual device from VS .NET using the Build menu, two .cab files are automatically copied to the device and extracted. Which .cab files are deployed is determined by the processor type and version of Windows CE running on the device. They include a development-only time .cab (`Sqlce.dev.platform.`*`processor`*`.cab`) that contains Query Analyzer and error string files, as well as the .cab file that contains the SQLCE database engine (`Sqlce.platform.`*`processor`*`.cab`).

When an application is ready for final deployment, the SQLCE .cab file must be added to the deployment and extracted on the device, as discussed in Chapter 10. The amount of space required on the device varies with the platform and processor, but it ranges from 1 to 3MB.

NOTE: In order to use SQLCE to connect to SQL Server 2000 using RDA or replication, additional configuration steps must be undertaken on the server machine as discussed in Chapter 7.

Deploying a SQLCE Database

Finally, it's important to note that in many instances it is more efficient and reduces load on the database server to prebuild a SQLCE database and deploy it to the device, rather than forcing clients to perform an initial synchronization using RDA or replication, as discussed in Chapter 7. This benefit only increases as the number of deployed devices in a solution increases. For example, a field service solution could be initially deployed with parts lists and geographic data.

To prebuild a SQLCE database, a developer can write an administrative application that creates the database on the device or the emulator and pulls in the appropriate data using RDA. The database can then be copied back to the development machine using ActiveSync and included in a VS .NET project as a content file using the Properties window. In this way, the database will be deployed with the application, as discussed in Chapter 10. Although it would be a welcome addition, at this time there is no desktop- or server-based utility to allow developers to create and populate SQLCE databases.

Alternatively, and especially if the database is large, the database file can be distributed on CompactFlash memory and CompactFlash disk drives, both of which are supported by SQLCE.

What's Ahead

This chapter has discussed the need for robust data caching and how that need is addressed using SQL Server CE. However, mobile applications that cache data locally using XML or SQLCE also typically need to synchronize their data with back-end systems. This final essential architectural concept will be addressed in the following two chapters, which look at both primitive synchronization using ActiveSync and more complex synchronization using RDA and merge replication.

Related Reading

Microsoft SQL Serve CE 2.0 Web site, at www.microsoft.com/sql/CE/default.asp.

SQL Server CE case studies, at www.microsoft.com/sql/ce/productinfo/casestudies.asp. Many of these case studies involve using RDA or merge replication or both.

Xue, Song. "SQL Server 2000 Windows CE Edition 2.0 Query Processor Overview and Performance Tuning Approaches." *Microsoft TechNet* (October 2002), at www.microsoft.com/technet/treeview/default.asp?url=/technet/prodtechnol/sql/maintain/Optimize/SSCEQPOP.asp.

Yao, Paul, and David Durant. "SQL Server CE: New Version Lets You Store and Update Data on Handheld Devices." *MSDN Magazine* (June 2001), at http://msdn.microsoft.com/library/default.asp?url=/library/en-us/dnsqlce/html/sqlce_secmodelscen20.asp.

Fox, Dan. *Teach Yourself ADO.NET in 21 Days*. Sams, 2002. ISBN 0-672-32386-9. See especially Chapter 18.

Gamma, Erich, et al. *Design Patterns*. Addison-Wesley, 1995. ISBN 0-201-63361-2. See p. 87 and following for a discussion of the Abstract Factory pattern.

Download the page for the Sybase Anywhere Studio .NET Data Provider from www.sybase.com.

Fox, Dan. "Protect Private Data with the Cryptography Namespaces of the desktop Framework." *MSDN Magazine* (June 2002), at http://msdn.microsoft.com/msdnmag/issues/02/06/Crypto/default.aspx.

CHAPTER 6

Primitive Synchronization

Executive Summary

In the previous chapters we've looked at the first three architectural concepts of local data handling, RDA, and robust data caching; however, applications require not only that data be accessible and transferred to a device, but that the transfer be intelligent and move data in both directions. This is the essence of the last architectural concept covered in this and the following chapter on synchronization.

More robust forms of synchronization using SQL Server 2000 Windows CE Edition will be covered in the next chapter. In this chapter we'll focus on using the features of ActiveSync, Microsoft's desktop synchronization software, to transfer file-based data to and from the device.

ActiveSync has evolved over the last several years, and its latest version, 3.7, includes the ability to perform backup and to restore, install software on the device, interact with the device graphically, perform file conversion, remotely communicate from a desktop application programmatically, and notify an application when connections are made. Because of the breadth of functionality and how it is exposed to developers, solutions can utilize ActiveSync to create a custom ActiveSync provider that piggybacks on the file-synchronization provider that ships with ActiveSync, develop a custom application that uses the remote API (RAPI) from the desktop, and create a custom application that takes advantage of the network pass-through functionality of ActiveSync 3.5 and higher.

When considering these options, creating an ActiveSync provider is the most difficult for Compact Framework developers because the Compact Framework does not support the COM Interop functionality needed to build an ActiveSync provider. Using file synchronization, however, allows a managed application to provide synchronization support with a modicum of effort. Although ActiveSync provides most of the infrastructure, this chapter will discuss a few additional nuances, including avoiding file collisions and implementing automatic notification. RAPI applications provide a more customized mechanism and rely on ActiveSync only for connectivity. The uses for RAPI applications are virtually unlimited due to the breadth of RAPI functions that allow remote activities, such as retrieving device system information, accessing the device registry, controlling processes, communicating with device windows with Windows messages, and performing directory and file manipulation, including the copying of files in either direction. Finally, pass-through applications can be written when using ActiveSync 3.5 and higher, coupled with Pocket PC 2002. In this scenario the PC acts as the hub to the network for the device and has the advantage of allowing synchronization directly with server software, allowing for more flexibility and scalability.

Regardless of which option is chosen, primitive synchronization like that discussed in this chapter typically works at the file level, and so, more granular requirements can be satisfied using SQL CE synchronization, as discussed in the next chapter.

The Importance of Synchronization

From the early chapters of the book, we have discussed mobile development from the perspective of data handling by exploring essential architectural concepts. In this journey, we have covered local data handling, remote data access, and caching data. If you combine the functionality of these three concepts into an application, you will be able to design software that pushes and pulls large amounts of data, regardless of whether it is already located on the device. In environments where the application functions over limited bandwidth connections, however, these techniques may be frustrating to the user due to the waiting involved. Therefore, the amount of data transmitted from constrained devices should be limited, and this brings us to the concept of synchronization and this chapter. Therefore, synchronization is the last essential architectural concept in our series of architectural points.

KEY POINT

Synchronization is distinct from the transferring and caching of data, which was thoroughly discussed in the previous chapters. Synchronization is important because it checks for the differences between two data containers in order to avoid the unneeded transfer of data that already resides in both data sources. Therefore, synchronization schemes typically update both data sources by transferring only additions, changes, and deletions.

Consider the scenario where a company has a database of customers. Inside the office, the customer service department executes a Windows application that allows agents to update the customer information located in the database. Also, salespeople have a Pocket PC application that allows them to download the customer database and edit customer records while out of the office.

The challenge in this application will be the synchronization of data from these traveling devices. If a developer were to develop this functionality, code would have to be written on the device and on the server. The effort to get this correct is tougher than you might initially think due to error handling and conflict resolution. Remember, the application cannot afford to lose data due to a loss of connectivity during transfer. In addition, a protocol has to be developed to query for the differences in data. Additions and deletions are easy to manage, but changes to existing data provide an opportunity for the developer.

The primary issue that synchronization needs to address occurs when a record is changed on both computers. For example, a customer table has an entry for Customer 100. The Pocket PC application downloads the customer database to the device that is taken out of the office by a salesman. During the day, the salesman discovers that Customer 100's phone number has changed and updates the customer record in his Pocket PC application. Later that day, someone in the customer service department notices the same discrepancy and makes a change to the database in the office. Which is the change that should remain? This is a simple case. But what would happen if these entries are different? Which is the right one? Should the application just accept the most recent one? What if the changes were made to the same customer but to different fields? The challenge here is how to manage conflicts.

There are several ways to achieve synchronization. It is certainly possible to create your own algorithm, which may consume significant effort based on the above scenario. Another alternative is to look at solutions built by others. Microsoft has built two mechanisms, ActiveSync and SQL Server 2000 Windows CE Edition. In this chapter, we will look at ActiveSync, and in the next chapter, we will look at the options provided SQL Server CE.

In exploring ActiveSync, we will not only look at how it does synchronization, but we will also look at the other services that it provides.

What ActiveSync Is

Any discussion of primitive synchronization and how it can be used when building solutions with the Compact Framework must begin with an understanding of ActiveSync.

As Microsoft made its foray into mobile devices, the company knew that Windows CE absolutely required the ability to exchange data with the PC. By connecting a cradle to a serial port and running special data-exchange software on both the PC and the device, Windows CE could intelligently retrieve data from the desktop to use in different applications on the device, thus giving the user a better experience with Microsoft's PDA. This software became known as ActiveSync.

ActiveSync is a lot of things today. If one is new to the PDA environment, it may appear that ActiveSync is just software that runs on the PC and intelligently transfers Outlook data, as in Figure 6-1; however, it is much more than this. ActiveSync creates a bridge between a desktop and a device that provides for a variety of services. Not only does it work across different connections (like USB, serial ports, infrared, and even via an Ethernet link across the network), it also works for Pocket PC, Handheld PCs, and the SmartPhone.

The most recent version of ActiveSync as of this writing is version 3.7 and is hosted on the desktop by WCESMgr.exe. For a better understanding, consider the services of ActiveSync outlined in the remainder of this section. These include backup and restore, software installation, interaction with the device, file conversion, remote communication, and connection notification.

Backup and Restore

ActiveSync provides the basic ability to backup and restore the data on the device. This feature allows a user to select where on the PC to store the file

Figure 6–1 *The Display of ActiveSync.* The typical view of ActiveSync from the desktop.

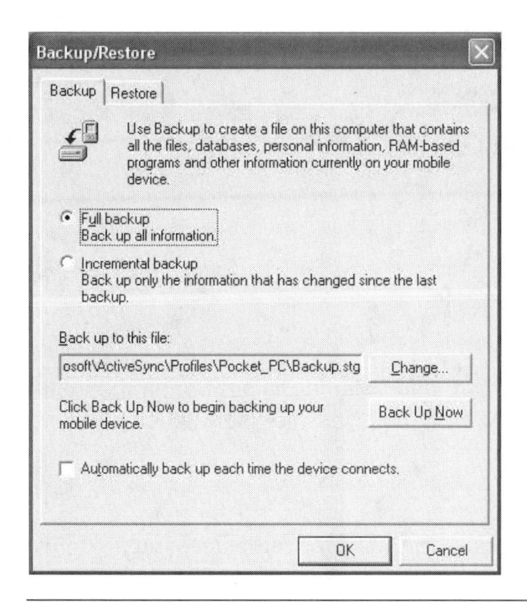

Figure 6–2 *ActiveSync's Backup/Restore Dialog*. From this screen shot, you can see that ActiveSync allows for full and incremental backups and provides for restoring the device.

(with a .stg extension), whether to do a full or incremental backup, and if the backup should occur on every synchronization of the device. In case of a device catastrophe, one would restore the backup file to a new device, reset the device, and then be glad for backups. Both of these options are on the Tools menu of the ActiveSync desktop software, as shown in Figure 6-2.

This feature can be combined with power-on password protection as described in Chapter 9 to form a simple means of data security. In this way, in the event of a misplaced or stolen device, the data can be restored onto a new device, while the old device cannot be accessed without removing the batteries and, therefore, totally wiping out its contents.

Software Install

ActiveSync also serves as a conduit for the device's application installation process. The most prevalent type of application installation for a device today is one that is initiated from the PC and is dependant on ActiveSync and its Application Manager. Through this mechanism, ActiveSync enhances the installation by prompting the user to provide the installation location on the device, which includes the choice of main memory or storage cards. The benefit to doing this is that the PC has a backup of the installation and that

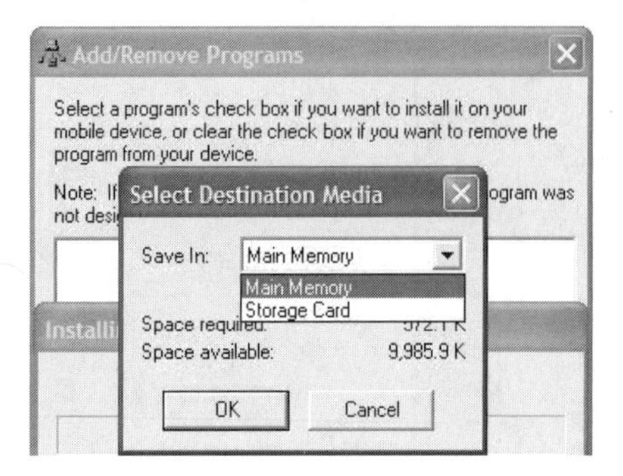

Figure 6–3 *ActiveSync's Destination Media Dialog*. During the install of a Windows CE application, ActiveSync gives the user a choice of installation locations.

it can later serve in reinstallation or removal if necessary. Figure 6-3 shows how the location of the installation can be chosen by the user.

In Chapter 10, Packaging and Deployment, an example is given that shows how to create this type of installation using the Application Manager (CeAppMgr.exe).

Mobile Explorer

ActiveSync also provides a nifty feature via the Windows Explorer on the desktop. After a synchronization, a user can browse or explore the device's storage system via the mobile device drive in Windows Explorer or the Explore button on the ActiveSync toolbar, just as if viewing the PC's hard drive as shown in Figure 6-4. This provides the ability to drag and drop different types of files to or from the device. Upon being dropped, a file can optionally be processed by a filter to perform the appropriate conversion for the destination (device or desktop).

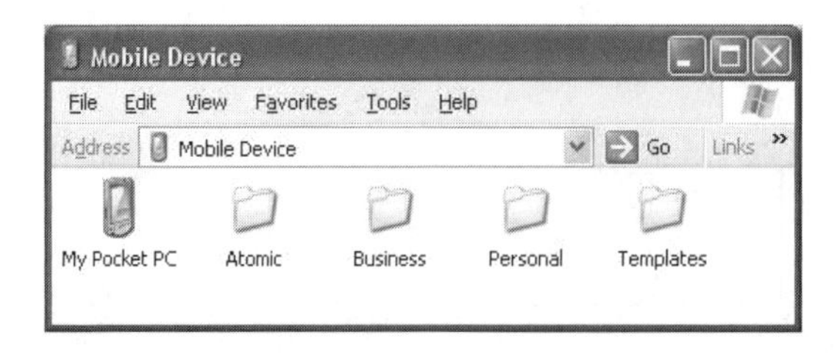

Figure 6–4 *Windows Explorer Browsing the Connected Device*. Due to ActiveSync, a connected device's file system is fully explorable.

File Conversion

As previously mentioned, ActiveSync can perform conversion operations on files as they are transferred between the desktop and the device. The code to do this is packaged as in-process COM objects installed on the desktop machine. Referred to as filters, these objects are intended to convert files as appropriate for the destination. In most cases, the files on the device are limited due to size constraints and reduced functionality in the application; therefore, the filters are important.

Because filters are COM objects and because the Compact Framework does not support COM Interop, as mentioned in Chapter 2, Compact Framework developers will not be able to create filters in Visual Studio .NET. As a result, developers must resort to eVC to do the COM programming. Fortunately, several important filters come with the current version of ActiveSync including those for Word, Excel, and bitmap files.

Remote Communication

The underlying service that makes the ActiveSync services discussed in this section possible is RAPI. Its purpose is to allow the desktop to execute Win32 API calls against the device. This functionality is located in Rapi.dll, an unmanaged DLL that resides on the desktop and includes 78 functions. Most of RAPI's functions are duplicates of Windows CE API functions on the device and are prefaced with a "Ce." For example, there is a Win32 function called `CreateProcess` that allows an application to start up another process on the device; therefore, the RAPI version is called `CeCreateProcess` and provides the ability to create a process on the device from the PC.

The functionality in RAPI is quite broad and allows for a variety of desktop-application purposes. Later in the chapter, we'll look at a sample that shows how to access RAPI from managed code.

Connection Notification

ActiveSync provides several ways to receive notification of connectivity, as well as other events. From the device side there are two techniques applications can employ. The first technique utilizes COM communications. The second is to register a program to be invoked when certain events occur. In this latter case, a developer could put an entry into the registry or call one of the Windows CE API functions (`CeSetUserNotificationEx` or the older `CeAppRunAtEvent`), about which more will be discussed later in the chapter.

The ActiveSync Architecture

As stated earlier, ActiveSync is concerned with the synchronization of data between the device and the desktop. It is interesting that the architecture is based on the concept of ActiveSync providers, so that a variety of synchronization types can be supported. When looking at this architecture, as Figure 6-5 shows, you'll notice that there are both desktop and device components and that both are based on COM.

In the remainder of this section, we'll briefly review the two primary components of the ActiveSync architecture, the Service Manager and ActiveSync providers.

ActiveSync Service Manager

The Service Manager is the core engine that executes the synchronization process. From Figure 6-5, you'll notice that it resides on both the device and desktop. It handles tasks such as establishing the connection between desktop and device, searching for changes in data by interacting with the registered providers (covered below), resolving conflicts based on rules configured in the ActiveSync client, and transferring the data. Acting as the manager of the synchronization process, the Service Manager does not decide which data to keep, delete, or transfer. These decisions are made in the ActiveSync providers.

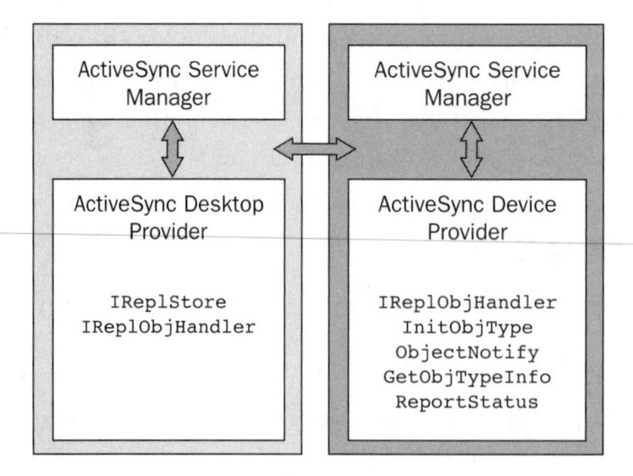

Figure 6–5 *The Architecture of ActiveSync.* This diagram illustrates the desktop and device software used by ActiveSync.

ActiveSync Providers

An ActiveSync provider includes COM objects packaged in DLLs that implement the `IReplObjHandler` and `IReplStore` interfaces and that invoke the Windows CE functions shown in Figure 6-5. Each provider is responsible for synchronizing a specific type of data. Once registered in the official list of ActiveSync providers via the desktop registry, the ActiveSync Service Manager will interact with the provider object via the COM interfaces. Several providers ship with the ActiveSync software, including the Microsoft Outlook ActiveSync provider which synchronizes Outlook data on the desktop with the various Pocket applications on the device, including the inbox, calendar, contacts, tasks, and notes. Others exist as well, including the files provider, which we will use later in this chapter.

As mentioned previously, the ActiveSync provider consists of a pair of DLLs, one for the desktop and one for the device. Each must implement the requisite COM interfaces in order to be an official provider. Because COM must be used to create an ActiveSync provider and because COM Interop is not supported in this release of the Compact Framework, managed developers will have to wait for a future version to implement an ActiveSync provider in VS .NET. Until that time comes, as with file conversion filters, eVC must be used.

Creating a Partnership

When a device is connected to a desktop running ActiveSync, the PC side of ActiveSync goes to work identifying the device. If the device's identity is not found, then the New Partnership dialog is displayed where the user is given an opportunity to choose between making a named partnership or synchronizing as a guest, as Figure 6-6 shows.

If the device is attached as a guest, then the ActiveSync functionality is limited. Guest capabilities include only backup and restore, browsing the device in Windows Explorer, accessing the network through the PC (known as ActiveSync pass-through), and software installation.

If a named partnership is formed, then the user is given an opportunity to select from the registered ActiveSync providers on the desktop. After selecting the providers, synchronization is performed and will thereafter be performed upon every connection (a partnership also has the guest mode functionality enabled). The provider selections and other synchronization options are stored in the registry in a key specific to the partnership. While the exact location of this information can be found by exploring the registry

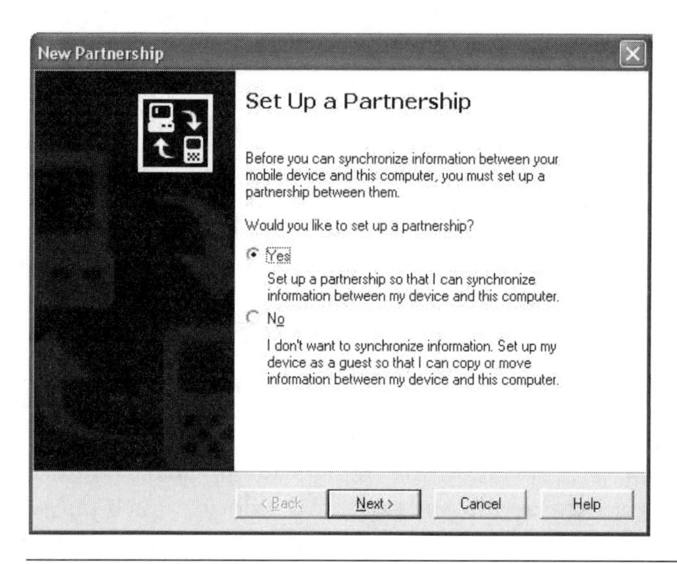

Figure 6–6 *New Partnership Dialog*. When the device does not have a partnership with the desktop, this dialog pops up upon cradling.

on the desktop and device, it is a best practice to use the CeUtil.dll, an unmanaged DLL (not a COM DLL). This DLL exposes functions such as `CeSvcOpenEx` and `CeSvcSetString`, which provide registry information specific to the device's locale and ActiveSync version (as these parameters will have a possible impact on partnership registry locations) and which can be called using the PInvoke functionality described in Chapter 11. With this, DLL developers can register desktop file filters and desktop synchronization partners, access device partnership settings that are used for both file filters and synchronization services, and add custom menu items.

Other issues to consider when using ActiveSync include the following:

- There is little documentation and support for using the emulator with ActiveSync; however, there is an article on MSDN that talks about how to do this using a serial cable and two ports on your PC.[1]
- ActiveSync does not natively synchronize between devices or with servers.
- For a device to interact with ActiveSync on the desktop, ActiveSync has to be running on the desktop, which requires a user to be logged in to the desktop with ActiveSync installed and running.

[1] This article is referenced in the "Related Reading" section at the end of this chapter.

- Partnership information is stored in the HKEY_CURRENT_USER registry hive, which means that the partnership is tied to a specific Windows account and the user's registry.
- A device can have a partnership with up to two desktops.
- Wireless synchronization cannot occur until a partnership has been formed via a cradle.

Developing ActiveSync Applications

Based on the functionality described thus far, several types of applications can be built based on ActiveSync services and used to synchronize device data with a desktop PC:

- *ActiveSync provider:* As described in the ActiveSync provider section, the ActiveSync architecture is extensible and allows custom providers to be implemented. Although this approach is the most difficult of those described in this section for Compact Framework developers (because of the reliance on COM), this type of application represents the tightest integration with ActiveSync.
- *File synchronization:* Although it is not typically discussed in most Windows CE resources, using the file-synchronization support included in ActiveSync allows a managed application to provide synchronization support with a modicum of effort; however, developers will likely want to add ActiveSync notification support, which will require a few calls to the Windows CE API. Applications that depend on file synchronization require an installed dummy file filter, which transfers files with a specific extension and stored in a special folder. This is discussed in more detail later in this section.
- *RAPI application:* This type of application resides on the desktop and calls functions located in the Rapi.dll. The application would therefore not interact with ActiveSync at all, and so the developer would have full control over the GUI. The uses for this type of application are virtually unlimited due to the breadth of RAPI functions that allow remote activities, such as retrieving device system information, accessing the device registry, controlling processes, communicating with device windows using Windows messages, and doing directory and file manipulation, including the copying of files in either direction. A simple example of how to use RAPI is included in this chapter.

- *Pass-through application:* When using ActiveSync 3.5 and higher coupled with Pocket PC 2002, the PC becomes the hub to the network for the device. To take advantage of this, no configuration is required. The developer simply builds the Compact Framework application to generate messages that access the network or Internet resources. These messages would "pass through" the PC and could be used with the types of communication that were covered in Chapter 4.

In the remaining parts of this section, we will discuss in more detail the latter three types of applications.

A Managed Application Using File Synchronization

Because this chapter emphasizes the most basic of synchronization options, we'll start by showing how the Compact Framework can be used to take advantage of the file-synchronization mechanism provided by ActiveSync. We'll illustrate this technique by synchronizing data for a list of book publishers between the desktop and device utilizing XML files generated from the database and an ADO.NET `DataSet` on the device. To do this, the following steps are required as discussed in the proceeding sections:

1. Enable file synchronization.
2. Create an ActiveSync dummy file filter.
3. Create a folder structure to support imported and exported files.
4. Build a PC application that generates files destined for transfer to the device and reads files transferred from the device.
5. Build a Compact Framework application registered to run upon a notification from ActiveSync.
6. Build a Compact Framework application that reads files transferred from the desktop and outputs files transferred to the desktop.

Enabling File Synchronization

The first step is to enable file synchronization from the desktop machine. To turn this feature on, a developer would select Files from the Options dialog (the same dialog shown during the partnership creation process) shown in Figure 6-7, invoked from the Tools | Options menu in ActiveSync.

Creating a Dummy File Filter

A dummy file filter is an ActiveSync configuration entry that tells the ActiveSync synchronization engine to transfer a file of a specific extension to the other

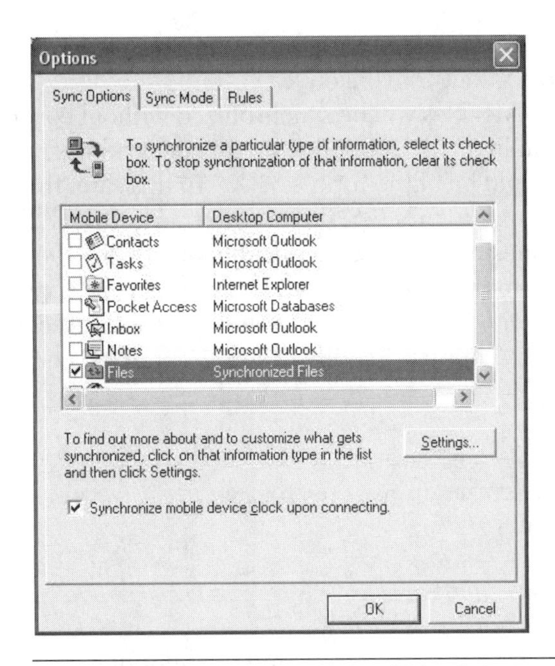

Figure 6–7 *The Options Dialog*. This dialog appears when creating a partnership. There are various Outlook features to choose from. File synchronization can be enabled by checking the Files checkbox.

side without conversion. By doing this, ActiveSync will copy data files not only from the desktop to the device, but also from the device to the desktop.

The filters for a machine are installed in the registry under the key `HKEY_LOCAL_MACHINE\Software\Microsoft\Windows CE Services\Filters`. As seen in RegEdit (Figure 6-8), ActiveSync installs a variety of filters. Each subkey is a file extension of a registered file type. Under the subkey, there are string values for `DefaultImport` and `DefaultExport`. The entries hold the class identifier (CLSID) of a COM object responsible for the conversion and for transferring data between the desktop and device. Note that

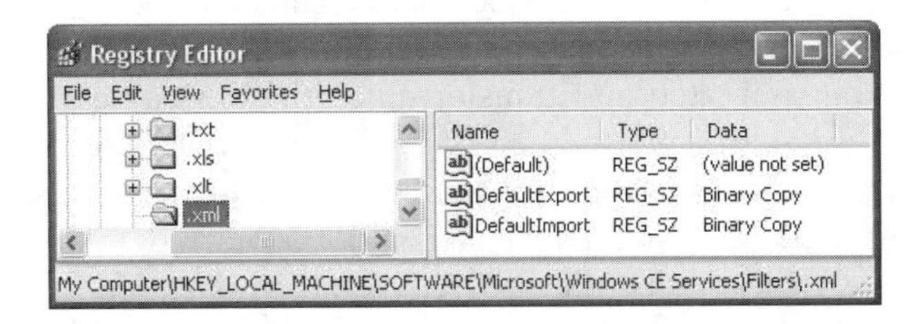

Figure 6–8 *RegEditLlooking at the Registered ActiveSync Filters.*

these entries can also have a value of `Binary Copy` to specify that the files should be transferred without conversion.

For our example, XML files will be synchronized without conversion for the application. To enable this, a developer would add a subkey to the registry key. The subkey should be of the form ".XXX." In this case, the extension would be .xml, as shown in Figure 6-8, in order to transfer XML files. Additionally, two string values are entered under the subkey for the `DefaultImport` and `DefaultExport`, which have the values `Binary Copy` assigned.

Although RegEdit can be used to make these entries, a second way to create these entries is to import a .reg file, the content of which is shown in the following snippet:

```
Windows Registry Editor Version 5.00
[HKEY_LOCAL_MACHINE\SOFTWARE\Microsoft\Windows CE
  Services\Filters\.xml]
"DefaultImport"="Binary Copy"
"DefaultExport"="Binary Copy"
```

Save this text file with a .reg extension, and then double-click to add the keys to the registry.

It is possible to view and edit the registry settings within the ActiveSync application by displaying the Options dialog, selecting Files, and selecting the Rules tab. From here a developer can disable the conversion of files at synchronization or edit the settings using the Device to Desktop and Desktop to Device tabs.

Creating a Folder Scheme

When a partnership is created, a synchronizing folder is created on the PC. The location of the folder is based on several factors, including where Windows user folders are stored, the Windows account name, and the name of the device. For example, if personal folders are stored in C:\Documents and Settings\, the user is jbox, and the device is named Pocket_PC, the resulting folder is C:\Documents and Settings\jbox\My Documents\Pocket_PC. Assuming that a filter is registered for a given file type, any file placed in this directory structure will be converted and transferred to the device's corresponding folder.

The device's synchronization folder will be \My Documents. Any file placed in this folder will be converted and transferred to the desktop machine (actually, the results will vary based on the registered filter).

In order to avoid conflict-resolution issues with using the same folder, we recommend creating an additional desktop folder with child folders for importing and exporting. So, in this example, we would have a folder called Publishers and two child folders called Inbox and Outbox. From the perspective of the device, Inbox is the container for files flowing to the device, whereas Outbox is for data flowing to the desktop. This structure should be placed under the file-synchronization folder as described in the previous paragraphs. This structure will then be automatically duplicated on the device during the next synchronization or immediately if the device is cradled.

This functionality can be tested by placing the files directly in the folders with Windows Explorer (as opposed to automating). For example, a developer could perform the following test to make sure everything is working:

1. Cradle a device with a partnership that has the synchronization setup, as we have discussed so far.
2. Find an XML file on the PC, and place it in one of the Publisher folders on the PC. In several seconds, the file will be on the device.
3. Delete the file from the PC. After several seconds, the file will disappear from the device.
4. On the PC, right click on one of the Publisher folders and create a text file. After several seconds, the file will show up on the device.
5. Click on the device file to load it in Pocket Word. Add some text like "From the Device," close the editor, and wait several seconds.
6. Open up the file on the PC to view the results.

Following the Data

Now that you have an understanding of the plumbing provided by ActiveSync, we'll discuss how both static and dynamic data will flow between the desktop and device.

The first flow is for static data. This is the type of data that rarely changes and usually comes in pairs (ID and description), for example, a list of states and their codes, customer types and their IDs, or other lookup and reference information that the device application will use to validate data. To move this type of data, the desktop application will need to query a database or other data store and write out the data at start-up and anytime the next flow occurs. By creating a Static.xml file in the Inbox folder, the file will be transferred to the device. The device application should notice this at some point, apply it intelligently, then delete the file from the device folder.

The second flow is for dynamic data, defined as any data that can be created, changed, or deleted on the device. This data must eventually be synchronized with the database or data store so that changes from any device will eventually find their way to the rest of the devices. This flow is initiated when the device is cradled. At this point, a notification application (using RAPI, for example) will send a message to the device application to initiate the task of writing all data to the Outbox (or launch the device application and write out the data), which will be synchronized to the desktop in a file called Dynamic.xml. This will be recognized by the desktop application, which will cause the data to be applied to the database, a full copy of the data to be generated from the database resulting in Dynamic.xml and Static.xml files in the Inbox folder (if needed), and the original Outbox/Dynamic.xml file to be deleted. The device will then notice the newly received files and start using them.

Building a Desktop Application

For this scheme to work, a process on the desktop has to push data from the data source and pull data from the device. Although it is a simple application to create, there are a few issues to consider.

The first is whether to build a Windows application or a Windows service. Fortunately, either type is easily built with VS .NET 2003.[2] Typically, a Windows service will be used if the situation requires a process that must run in the background, doesn't require user interaction, and should be running without someone initiating the process. Although a Windows service sounds like the right solution for this scenario, regrettably, ActiveSync requires a logged-in user; however, it still has the advantage of not requiring a user to initiate the process or worry about who is logged in. These will still be issues for ActiveSync itself.

A second issue entails dealing with the folder names and their hard coding. The preferred technique to avoid hard-coding paths in the application is to use the managed application's configuration file. By storing the path in the configuration file, the programmer can use the `ConfigurationSettings.AppSettings` class to retrieve the data easily.

The final issue centers on how to deal with the data. This program is responsible for interacting with the database and the synchronization folders. This responsibility requires two activities: (1) writing out a copy of static data to Publishers\Inbox, and (2) whenever a file shows up in the Outbox, reading the data and updating the database, making a new copy of the

---

[2] See the article referenced in the "Related Reading" section at the end of the chapter for the basics on creating Windows services.

dynamic and static data in the Inbox, and then deleting the file in the Outbox. In order to be notified when a file reaches the Outbox, the desktop application can take advantage of the `System.IO.FileSystemWatcher` to avoid the polling logic using timers.[3]

By letting the device react to the cradling event and then pushing the dynamic data to the desktop, the process ensures that all updates are applied to the database before a new copy is sent to the device. Another benefit is that when the application is brought up for the first time, the static data will be there for start-up.

Creating a Device Notification Application

The device application must be launched by the operating system when the device is cradled. There are several ways to accomplish this, but we'll describe the registration route (because the other is COM based). This task is accomplished by creating a stub application that calls the Windows CE API functions `CeRunAppAtEvent` or `CeSetUserNotificationEx`. Chapter 11 includes an example of `CeRunAppAtEvent`, and detailed instructions for building a full example with `CeSetUserNotificationEx` are provided in the lab at the Atomic Mobility site (http://atomic.quilogy.com/mobility). This lab was used at TechEd 2003 and is the basis of this notification application. This code takes advantage of PInvoke, although interaction with it is encapsulated in classes to make it easier to utilize.

When this program is launched by the act of cradling, the first step is to inform the device application. This is done using the following steps:

1. Check if the device application is running by checking for a mutex that the device program has created.
2. If the mutex exists, use the `MessageWindow` class to send a message, and then exit.
3. If the mutex does not exist, launch the application with command-line parameters to start a refresh, and exit the program.

Although you would think it would be possible to include this logic in the device application (described below) instead of having two applications on the device, the Compact Framework runtime will not allow a second instance of an application to run on the Pocket PC.

[3] See Chapter 12 of Dan Fox's *Building Distributed Applications in Visual Basic .NET* referenced in the "Related Reading" section at the end of the chapter.

Using a Device Application

The device application is the Compact Framework application that the disconnected worker is to use. The application will likely utilize a `DataSet` as described in Chapter 3 for its local data storage. Other than providing the user with the UI for viewing and maintaining the data, the application will have several additional responsibilities related to the synchronization tasks.

For first time execution, the application should send an empty data set to the Outbox, which will launch the synchronization process at the desktop, resulting in a copy of the `DataSet` back at the device. If necessary, the desktop will create in the Inbox a Static.xml file that contains necessary lookup information for the application. During the time that the application is waiting for the data to return from the desktop, no changes to existing data should be allowed. When new files appear in the Inbox, the application should start immediately using the new data.

At this point, anytime a refresh message is received from the device notification application, the device application should create a Dynamic.xml file in the Outbox and wait for a Dynamic.xml file to return to the Inbox. Upon return, the dynamic file will act as the database for the device. This process is simply a subset of the initialization process, and so there is an opportunity here for code reuse.

When using this type of architecture, the Compact Framework will take advantage of the capability in ADO.NET to serialize and deserialize ADO.NET `DataSet` objects as XML files, as described in Chapter 3; however, be aware that the larger the XML files, the more time start up will require.

Conclusions on File-Synchronization Application

To summarize the process described in the preceding sections, consider the full process of using file synchronization as outlined in Figure 6-9 and described here.

1. Start the device notification application on the device. It does one of two things, depending on if the device application is running (determined by checking for an existing mutex): (a) If the device application is running, it does a broadcast using `MessageWindow` class, or (b) if the device application is not running, it does a `CreateProcess` on the device application passing command-line parameters to indicate an immediate refresh is needed.
2. Write out Dynamic.xml to Outbox.
3. ActiveSync replicates the file to the desktop Outbox.

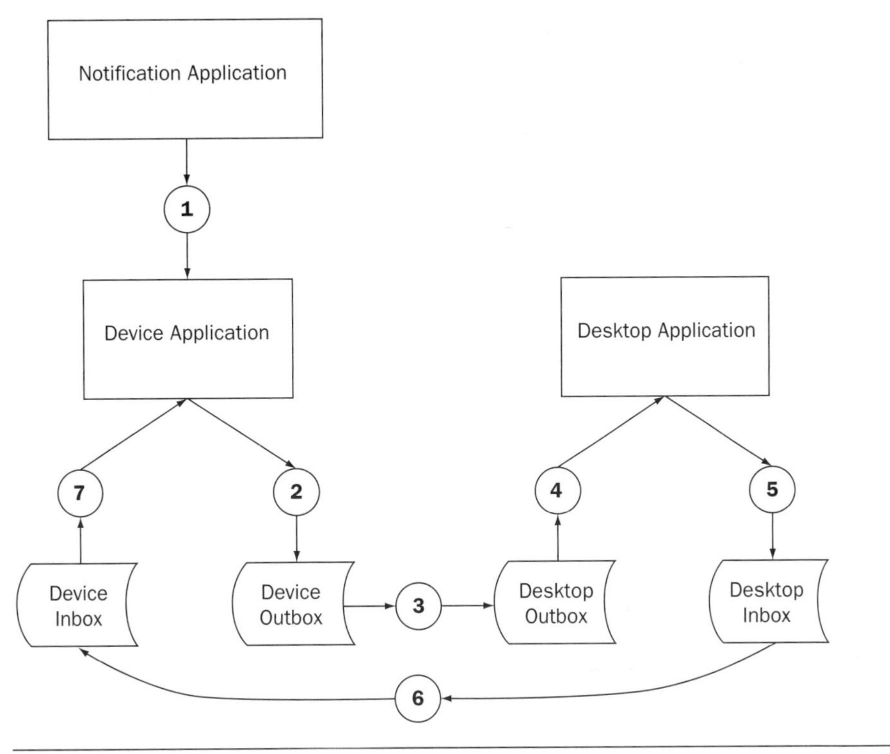

Figure 6–9 *The File-Synchronization Process.*

4. The desktop application recognizes a new file arriving (for example, by using the `FileSystemWatcher`). The application reads the data and applies the data to the data source. The Inbox file should be deleted upon reading, which will delete the copy on the device.
5. The desktop application writes out Static.xml and a Dynamic.xml files to the Inbox folder. The Static.xml file includes a fresh copy of lookup data, and Dynamic.xml incorporates the changes from other users.
6. ActiveSync copies the files to the device Inbox folder.
7. The device application has been suspended since step 2. It now sees a file in the Inbox folder. It reads the files as `DataSet` objects, deletes the files from the Inbox (which deletes them from the desktop), and allows normal operation to continue.

File synchronization seems fairly simple at first glance; however, it does entail issues you and your team will need to consider. First, ActiveSync forces the device to be tied to a desktop machine. Second, scalability is limited due to working with a desktop process instead of a server process. Third, file synchronization copies entire sets of data, whereas a more robust synchronization process will allow for more granularity by allowing users to drill down to the record or field level. For these sorts of scenarios you'll want to explore the SQL Server CE synchronization discussed in the next chapter.

NOTE: The primary scenario for using file synchronization is when the master data source will be constrained to the desktop machine.

If the device does not have network connectivity, then cradling is still required; and although you might think that this automatically means that file synchronization must be used, you should also explore the ActiveSync pass-through capability described later in this chapter.

Utilizing RAPI in a Managed Application

As mentioned previously, RAPI is a core part of ActiveSync and allows a desktop application to control and query a device. Due to the need to access Rapi.dll, a .NET Framework application can use the PInvoke service described in Chapter 11. To give you a sense of how RAPI can be utilized, consider Listing 6-1, a simple example in C# that launches a game of Solitaire on the device from the desktop machine.

Listing 6–1 *A RAPI Console.* This C# snippet uses RAPI to start up Solitare on the device from the desktop.

```
using System.Runtime.InteropServices;

[DllImport("rapi.DLL", CharSet=CharSet.Unicode)]
public static extern int CeRapiInit();

[DllImport("rapi.DLL", CharSet=CharSet.Unicode)]
public static extern int CeRapiUninit();

[StructLayout(LayoutKind.Sequential,Pack=4)]
public struct ProcessInfo
{
  public IntPtr hProcess;
  public IntPtr hThread;
  public int dwProcessId;
  public int dwThreadId;
};
```

```
[DllImport("rapi.dll", CharSet=CharSet.Unicode)]
public static extern
int CeCreateProcess (string lpApplicationName,
  string lpCommandLine, int Res1, int Res2, int Res3,
  int dwCreationFlags, int Res4, int Res5, int Res6,
  ref ProcessInfo lpProcessInformation);

[DllImport("rapi.dll", CharSet=CharSet.Unicode)]
public static extern int CeCloseHandle(IntPtr Handle);

static void Main(string[] args)
{
  // Intialize RAPI
  int hr = CeRapiInit();

  //if safe to continue
  if (hr==0)
  {
    string strProg = @"\windows\solitare.exe";

    //build needed structure used in API call
    ProcessInfo pi = new ProcessInfo();

    //make important RAPI call to start Solitare
    CeCreateProcess(strProg,  "", 0, 0, 0, 0, 0, 0, 0, ref pi);

    //release handles created process and thread
    CeCloseHandle(pi.hProcess);
    CeCloseHandle(pi.hThread);
  }

  // Shutdown RAPI connection
  CeRapiUninit();

  return;
}
```

NOTE: As with any calls to PInvoke, the `CeRapiInit`, `CeRapiUnit`, and `CeCreateProcess` functions can be encapsulated in a managed class and exposed as static methods to make them easier to call and maintain.

For more detailed information on using RAPI in Compact Framework applications, we recommend Chapter 16 of Paul Yao and David Durant's *Programming the .NET Compact Framework in C#,* as noted in the "Related Reading" section at the end of the chapter. Not only does this chapter provide coverage of the 78 functions in RAPI, it also explores more advanced examples, including how to interoperate with COM.

Considerations for a Pass-Through Application

Even though the synchronization example looks interesting, it should be used only in specific scenarios. Although some will suggest that a solution should use file synchronization when over-the-air security is an issue or when there is no network connectivity built into the device, there is another option as well.

These issues are also good reasons for using the cradle, but for not using file synchronization. A more robust architecture is one that uses servers, rather than desktops; therefore, if the requirements include cradling, the solution can still utilize servers. One of the primary advantages, of course, is that scalability is more achievable, for example, when using SQL Server as opposed to Access or MSDE on a desktop.

Fortunately, ActiveSync can assist this type of application as well. With the combination of ActiveSync 3.5 (and higher) and Pocket PC 2002 (and higher), ActiveSync provides network connectivity to a device while cradled. This is known as ActiveSync pass-through and gives a device access to the network accessible from the desktop PC using TCP/IP. As a result, it is possible to have a cradled device that communicates directly with servers on a network.

There is, however, the issue of partnerships to resolve. In the default configuration of ActiveSync, the cradling of an unrecognized device always creates a prompt on the desktop asking users to choose to create a partnership or act as a guest, as described previously. In many basic scenarios the mobile application requires using PCs that have cradles, and those PCs are often located in field offices that may already be logged in under a different user's identity. As a result, either the logged on user will have to logoff and the roving user login, or the logged on user will have to create a partnership or choose guest with each of the devices that connect using its cradle.[4]

[4] Creating named partnerships is obviously problematic due to the existence of the PC user's Outlook information. You wouldn't want this information ending up on each device that synchronizes with the cradle on the PC.

To avoid forcing the desktop user to choose guest continually when synchronizing, there is a registry setting that will change the default behavior. By adding GuestOnly, a DWORD entry with a value of 1 that resides at HKLM\Software\Microsoft\Windows CE Services, the prompting will cease. Instead, sessions with unrecognized devices will automatically connect as guest. And, in these cases, the pass-through capability is all that is required.

What's Ahead

In this chapter we looked at primitive synchronization techniques that can be employed in Compact Framework solutions. These solutions necessarily revolve around ActiveSync, although, as you've learned, there are several options and issues to consider.

Because the synchronization techniques discussed in this chapter are necessarily limited, the next chapter focuses on the more advanced synchronization capabilities of SQL Server 2000 Windows CE Edition. And as you'll no doubt discover, in some cases the concepts from this chapter can also be combined with those in the next.

Related Reading

Boling, Douglas. *Programming Microsoft Windows CE .NET*. Microsoft Press, 2003. ISBN 0-735-6188-44. See Chapter 15, "Connecting to the Desktop."

Fox, Dan. *Building Distributed Applications with Visual Basic .NET*. Sams, 2001. ISBN 0-672-32130-0. See pp. 568–573 for information on encapsulating the `FileSystemWatcher` component.

Yao, Paul, and David Durant. *.NET Compact Framework Programming with C#*. To be published by Addison-Wesley. See especially Chapter 16 on RAPI and Chapter 21 on ActiveSync.

Wigley, Andy, and Stephen Wheelwright. *Microsoft .NET Compact Framework, Core Reference*. Microsoft Press, 2003. ISBN 0-7356-1725-2. See Chapter 13, "Accessing Data."

"Pocket PC Developers Guide: Using ActiveSync Over Serial," at http://msdn.microsoft.com/library/en-us/guide_ppc/htm/ppc_using_emulator_with_activesync__zxtb.asp.

Fox, Dan. ".NET Makes Windows Services a Breeze." October 2002, at www.informit.com/content/index.asp?product_id={79F04B5B-2167-41D3-8EFC-B8F6DB717261}

CHAPTER 7

Data Synchronization

Executive Summary

This chapter is a continuation of the discussion in the previous chapter covering the essential architectural concept of synchronization. Whereas in the previous chapter the focus was on synchronizing file-based and XML data, this chapter focuses on the two built-in synchronization mechanisms of SQL Server 2000 Windows CE Edition 2.0: RDA and merge replication. The obvious advantage to using these mechanisms is that the connectivity architecture and infrastructure work have already been done, thereby easing the work of architects and developers to integrate this functionality into custom solutions.

The connectivity architecture is the same for both RDA and replication and involves a client agent on the device making HTTP(S) requests to a server agent running on an IIS server. The server agent then connects to SQL Server and either executes commands, synchronizes replicated data, or returns rows to the client agent for processing. The client agent API is exposed through the `SqlCeRemoteDataAccess` and `SqlCeReplication` classes in the SqlServerCe .NET Data Provider so that Compact Framework applications can initiate requests. On the server side the configuration of IIS is managed by the SQL Server CE Connectivity Management MMC snap-in that administrators can use to configure the IIS virtual directory and its associated security. This architecture is attractive because of its accessibility via HTTP, security using IIS and SSL, performance through its use of a compression algorithm to transmit data, and reliability because the underlying block-based protocol includes automatic restarts at the last successful block.

RDA can be thought of as the more lightweight of the two mechanisms because it does not require any special configurations on the SQL Server. It can be used to pull rows on a per-table basis to the device, optionally track changes made on the device, and then later push those changes back to the server using a simple optimistic concurrency-based model. Error rows are then saved on the device for later review. For this reason, RDA is ideal for loading data that will be both dynamic and static, capturing data (where the data is created on the device such as through a bar code reader), and dealing with data that is highly partitioned, where it is unlikely that two users will update the same row. RDA can also be used in a connected mode to submit any SQL statement to the server (as long as it does not return rows). The API exposed by the `SqlCeRemoteDataAccess` class is fairly straightforward, but developers may wish to wrap some of its functionality, especially setting its connectivity and security properties, into a Singleton class.

Merge replication allows a SQLCE client to act as a subscriber to a merge publication created on SQL Server 2000. While this mechanism requires more server configuration and planning, it allows more control as well through server-based row and column filtering (including dynamic filters), the ability to load and synchronize multiple tables, server-based conflict detection and resolution, and bidirectional data flow, whereby changes made on the server also show up on the device. Together, these make replication ideal for occasionally connected applications with shared-data scenarios and where data must be incrementally updated on the device. The `SqlCeReplication` class exposes a simple set of methods for creating subscriptions and synchronizing data.

SQLCE Synchronization

In the previous chapter the discussion focused on the issues surrounding, and techniques involving, simple file-based synchronization using ActiveSync. In those scenarios for which it makes sense (typically where the application can use XML or other file formats to exchange data and the device will be regularly cradled for synchronization), using ActiveSync can be a very effective means of synchronizing data.

However, there are a number of other scenarios in which the application requires more robust data storage, as discussed in Chapter 5, and where the device needs to synchronize when not connected to a specific PC. In these scenarios, which typically involve a large number of devices using data from a single back-end data store (for example, a field-service automation application, where all the drivers in an organization download delivery and route information, and then later synchronize updates to the corporate database), the devices connect instead using 802.11 on a WLAN or GPRS over a WAN. For these applications, a more sophisticated synchronization process is required.

Fortunately for architects and developers, SQLCE includes two mechanisms for connecting to and exchanging data with back-end systems. These mechanisms, RDA and merge replication, also have support built in to the SqlServerCe .NET Data Provider, which makes it easy for those using the Compact Framework to take advantage of this prebuilt infrastructure. After a brief overview of the connectivity architecture and configuration of SQLCE, this chapter will focus on these two methods of synchronization to round out our discussion of the final essential architectural concept of synchronization.

Connectivity Architecture

Before applications can utilize the synchronization features of SQLCE, the SQL Server CE Server Tools shipped with VS .NET 2003 and included in the Compact Framework SDK directory (for example, sqlce20sql2ksp1.exe and sqlce20sql2ksp2.exe[1]) must be installed on a server running IIS.[2]

The server tools are required because they include the SQLCE server agent shown in Figure 5-2. It is the job of the server agent[3] to process HTTP(S) requests from the SQLCE client agent to IIS. The server agent then connects to SQL Server and either executes commands or returns rows to the client agent for processing.

NOTE: It is important to note that the IIS computer configured with the SQLCE server agent needn't and usually won't reside on the same computer as SQL Server or the back-end database. This allows for looser coupling and the ability to use the Windows Network Load-Balancing Service (NLB) and Application Center 2000 to create a load-balanced cluster (Web farm) of servers that process client agent requests from multiple devices. Various topologies are discussed at the end of this chapter.

In order to allow the client agent to make requests of the server, one or more virtual directories in IIS must be configured with the server agent to accept requests. This is accomplished using the SQL Server CE Connectivity Management MMC snap-in shown in Figure 7-1. Using this utility, an

[1] A version for service pack 3 of SQL Server 2000 is also available on MSDN at www.microsoft.com/sql/downloads/ce/sp3.asp.

[2] See the SQL Server CE Books Online for more information.

[3] This is actually implemented as an ISAPI DLL (Sscesa20.dll) that processes the HTTP request through IIS.

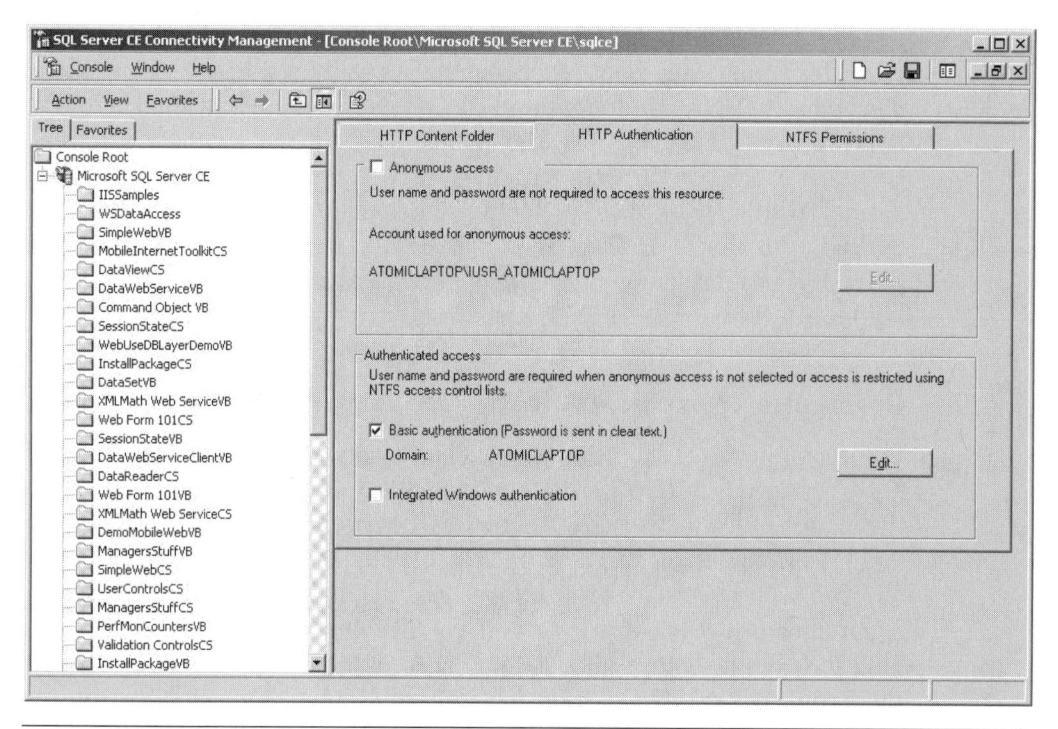

Figure 7-1 *The SQLCE Connectivity Management Snap-In.* This utility is used to configure a virtual directory for use with the SQLCE Server server agent. It can also be used to set HTTP and NTFS permissions, as in this case, where basic HTTP authentication is required.

administrator not only can specify the virtual directory, but can additionally configure the level of authentication required (anonymous, basic, or integrated Windows) by IIS on the virtual directory, additional NTFS permissions on the directory, and the NTFS permissions required on the share pointing to the snapshot folder used in merge replication.[4] For example, if basic authentication is configured, as shown in Figure 7-1, an administrator must also add the appropriate NTFS permissions for valid accounts (groups or users) using the NTFS Permissions tab.

To make it easier to configure and when an administrator needs to create a new virtual directory to use with SQLCE, the Connectivity Management snap-in also includes the SQL Server CE Virtual Directory Creation

[4] The snapshot folder is a folder on the SQL Server computer containing the schema and data for tables published through merge replication.

Wizard. This wizard can be accessed immediately when the Server Tools are installed or by double-clicking the icon in the Connectivity Management snap-in.

Connectivity Features

Several advantages to the SQLCE connectivity architecture just described include the following:

- *Accessibility:* Because the communication protocol used is HTTP over TCP/IP, devices can make requests for data that pass through firewalls. In addition, connectivity is supported through WLANs, WANs, and even a pass-through mechanism, using a directly connected device (serial, infrared, USB, or Ethernet) with ActiveSync 3.5 and higher or SQL Server CE Relay.
- *Security:* Because SQLCE connectivity uses IIS for its communication mechanism, the IP address and domain name restrictions, encryption of the communication using Secure Sockets Layer (SSL), and the authentication methods in IIS (except for digest, Kerberos, and client certificates not supported by Windows CE 3.0) can be used. This also implies that credentials can be authenticated against a Windows NT domain or Active Directory.
- *Performance:* In order to support wireless devices in an optimal way, both RDA and replication use compression when transmitting data between the client and server agents. This compression is on the order of 8:1 and serves to greatly improve the performance of applications exchanging large amounts of data. This is contrasted with communication to a back-end SQL Server using the SqlClient .NET Data Provider, which is not compressed.
- *Reliability:* The simple protocol used by SQLCE connectivity[5] can recover from communication failures of approximately two minutes by restarting from the last successfully transferred block of data. This allows synchronization to occur even when the underlying communication transport is erratic.

In addition to these primary advantages, of course, is that developers within your own organization don't have to reinvent the wheel to provide these basic infrastructure requirements.

[5] The protocol is patterned after various file-transfer protocols.

RDA

The first of the two synchronization mechanisms made available through SQLCE is RDA. This mechanism provides simple client push-and-pull functionality along with the ability to submit statements to a back-end data store using OLE DB. In a nutshell, the client agent provides an API that applications can use to submit requests over HTTP to the server agent. Compact Framework developers access this API through the `SqlCeRemoteDataAccess` class exposed by the SqlServerCe .NET Data Provider.[6] The server agent then uses a connection string sent from the client to initiate a connection to SQL Server 7.0 (service pack 4 or later) or 2000 and forwards the request to the data store as shown in Figure 7-2. The request is then processed by the SQL Server, and the results (rows, errors, messages), if any, are sent back to the device by the server agent. The SQLCE database engine manages the rows returned from an RDA pull request and can even track which rows in the table have been pulled from the remote data source so that they can be sent back during a push request.

In the remainder of this section, the scenarios applicable to RDA will be discussed, along with an overview of RDA security and examples of how it can be used in a Compact Framework solution.

Features and Scenarios

RDA includes several features that make it attractive as a part of a solution:

- *Reduced management:* Unlike merge replication discussed later in this chapter, RDA requires no special configurations on the back-end server. For example, a publication needn't be created on SQL Server 2000 in order for rows to be pulled down to a device, updated, and later pushed back to the server using RDA. This lessens the administrative burden for applications using RDA.
- *Simplicity:* Because connectivity to the data store is managed by the server agent and the SQLCE database engine tracks pulled rows, the API for using RDA is extremely simple and consists of just seven properties, three methods, and two enumerations in the Compact Framework, making it very easy to implement. In fact, when making

[6] eMbedded Visual Tools developers can use RDA through the RDA COM object by referencing the SQL Server CE Control 2.0 library in eVB or by including the appropriate files in an eVC++ project.

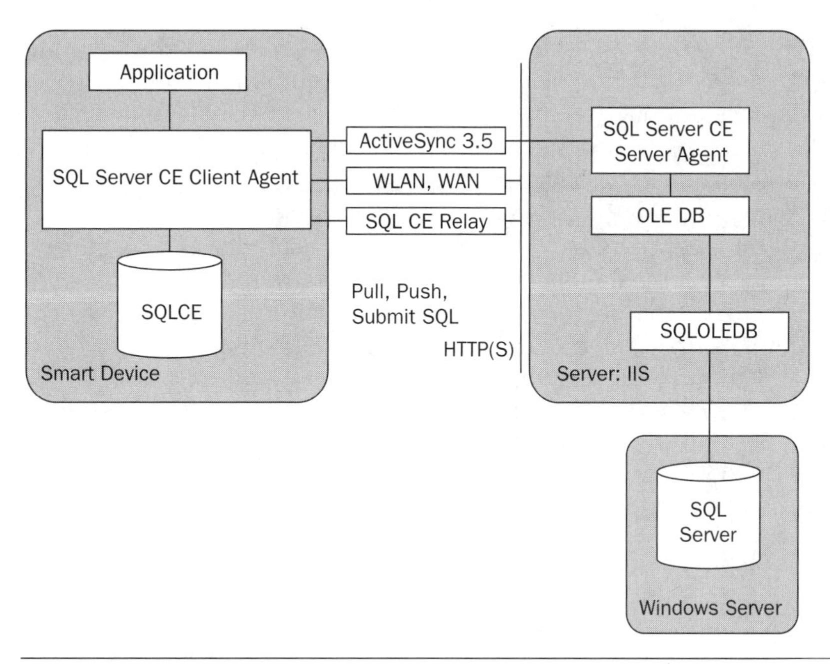

Figure 7–2 *RDA Architecture*. This diagram shows the architecture of RDA and highlights the fact that it is used with SQL Server 2000.

a pull request, the target table in SQLCE is automatically created, and RDA can even create the appropriate index on the table when pulled from SQL Server 2000.

- *Scalability:* Because there is no extra processing (change tracking and conflict detection and resolution) on the server side, as there is with merge replication, RDA provides for better scalability.

However, while RDA provides certain advantages, it is more limited than a full merge-replication solution in three specific ways:

1. Although RDA supports the ability to pull a set of rows from a data store and push modified rows back to a data store, it provides only client-side tracking. In other words, as other users make changes to data in a database, the client using RDA will not be aware of which rows have changed and will need to repull the set in order to see the changes. This is contrasted with merge replication where the server tracks changed rows and can send only the changed rows that the client doesn't already have.

Why Only SQL Server?

At first glance, it would seem that RDA would be an ideal technology for allowing SQLCE to interact with database products other than SQL Server; however, it's important to keep in mind that the SQL server agent running on the IIS machine does have to do some work to move data into and out of SQL Server, including mapping the data types between systems when creating SQLCE tables and generating SQL statements as changes are pushed from the device.

As a result, the server agent would need to include special handlers for each OLE DB data source that RDA could use, a task that Microsoft has not embarked on at the time of this writing.

To see how the server agent handles the mapping for SQL Server, see the "Supported Data Types and Data Type Mappings" topic in the SQLCE Books Online.

2. Closely related to the first item, RDA can support only a limited form of conflict resolution. Basically, as described later in the chapter, RDA uses optimistic concurrency where rows pulled to a device are not locked, and so, other users may modify them on the server. Because the server keeps no record of which rows were pulled, rows pushed from the device may overwrite these changed rows. RDA detects a conflict only if an error occurs while an attempt is made to insert, update, or delete the row.
3. The pull mechanism in RDA is limited to pulling data into a single SQLCE table at a time. Although the request sent to the data store can use `JOIN` clauses and even encapsulate a request to a stored procedure, the rows returned are placed into a single table. This means that developers who must work with multiple tables will need to formulate separate requests for each table. This is contrasted with replication, where a smart device application can subscribe to an entire publication, which may include multiple tables.

After boiling down the features and limitations of RDA, it becomes apparent that RDA is most useful in the following scenarios:

- *Data loading:* Because RDA pulls data from SQL Server and automatically creates the appropriate tables in SQLCE, complete with indexes, it is obviously useful for populating a SQLCE database with tables used for lookups in an application such as a price list or product information list. This might occur the first time a user starts a particular application or at predefined intervals in order to refresh a relatively static set of data. As mentioned in Chapter 5, RDA would also be appropriate for prebuilding SQLCE databases that are later deployed to devices via a LAN/WAN or storage device.

- *Data capture:* Some smart device applications are used simply to collect data that has no corresponding row on the server. For these scenarios, an empty row set can initially be pulled from the data store in order to set up tracking in the SQLCE table,[7] and then periodically, the new rows can be pushed back to SQL Server using RDA. In these sorts of scenarios, sometimes the data ultimately does not end up in a SQL Server, but in Oracle or DB2. For these scenarios you can think of the SQL Server as an intermediate data collection point from which you can then extract the data using SQL Server's Data Transformation Services (DTS), linked tables, or heterogeneous replication[8] to load into the main repository.
- *Working with highly partitioned data:* Because RDA does not include sophisticated conflict resolution, it works well in scenarios where data is naturally partitioned and few conflicts are expected. For example, a set of clients assigned to a particular account manager could be pulled to the device without fear of other users updating the client exchanges before the account manager pushes the data back to the data store.
- *Connected data manipulation:* In addition to the push-and-pull functionality, like the SqlClient .NET Data Provider discussed in Chapter 4, RDA supports the ability to send any statement to a data store that does not return rows. This allows an application to make modifications directly to the data store at any time and not have to wait for a specific synchronization. This could be used on a shop floor, for example, where the device is always connected to a WLAN; however, in practice, the SqlClient provider would likely be the first choice for such a scenario.

TIP: It is important to remember that because RDA requires that the table not exist in SQLCE when a pull is done, RDA is not well suited for applications that require incremental updates because RDA would need to pull down the entire table again, rather than simply the new rows.

---

[7] RDA requires that for a table to be pushed back to a data store, it must have originally been created through a `Pull` request so that tracking in the SQLCE engine could be enabled.

[8] Heterogenous replication is a replication option in SQL Server 7.0 and 2000 that allows any OLE DB data source to act as a subscriber to a SQL Server publication. See the SQL Server online help for more information.

Configuration

As mentioned previously, in order for either RDA or merge replication to work, the server agent must be installed on IIS, and one or more virtual directories would have to be created. Once accomplished, there are two other configuration issues to consider: security and logging.

Security

In order for requests initiated on the device by the client agent to access data successfully on the SQL Server, authentication and authorization must take place. The three layers of authentication that are supported include authenticating through a proxy server, to the IIS computer, and finally to SQL Server itself.

- *Proxy server:* The credentials for the proxy server can be specified using the `InternetProxyServer`, `InternetProxyLogin`, and `InternetProxyPassword` properties of the `SqlCeRemoteDataAccess` object. A proxy server or firewall cannot be used when integrated Windows authentication is used in IIS; so, integrated Windows authentication is most often used in intranet scenarios when the device will connect via a LAN. The `InternetProxyServer` property must also be used when using SQL Server CE Relay to connect to SQL Server via older versions of ActiveSync.[9]
- *IIS:* The location of the IIS server and its credentials for authentication are specified using the `InternetURL`, `InternetLogin`, and `InternetPassword` properties of the `SqlCeRemoteDataAccess` object. These are used when IIS is configured for either basic or integrated Windows authentication; however, as with typical Web applications, basic authentication will send the password as cleartext, and so SSL is recommended to encrypt the entire channel during communication. To use SSL, a server certificate must be installed on the configured virtual directory in the same fashion as on any other Web site. Windows CE maintains a database of trusted certificate authorities (CA) and consults this database when SSL connections are made. If your organization uses its own CA, you must perform a one-time update of the database of CAs on each device with your own root certificate, as

[9] In this case, the property is set to `ppp_peer:`*<client port number>*.

described in the SQLCE Books Online.[10] The IIS credentials are not required if anonymous authentication is used. As mentioned previously, regardless of the authentication mode in use, the account that is authenticated (either the user's account or the group, if using basic or integrated Windows, or the `IUSR_`*`computername`* account) must also be granted read and execute NTFS permissions to the server agent DLL in the virtual directory and read and write permissions to the folder itself, which can be done with the SQL Server CE Connectivity Management MMC snap-in shown in Figure 7-1.

- *SQL Server:* The SQL Server credentials can be specified in the connection string (formulated as an OLE DB connection string) passed to the `Push`, `Pull`, and `SubmitSQL` methods of the `SqlCeRemoteDataAccess` object using the `user id` and `password` properties, if SQL Server authentication is used, and the `Integrated Security=SSPI` setting, if integrated Windows authentication is used. When using integrated Windows authentication in SQL Server, however, the token that IIS will send to SQL Server depends on the IIS authentication mode. For example, if anonymous is activated, then the guest account's token configured in IIS (typically *`computername`*`\IUSR_`*`computername`*) is used. If basic or integrated Windows authentication is configured in IIS, then a token created with the `InternetLogin` specified on the device is passed to SQL Server. The caveat is that the SQL Server must reside on the same machine as IIS if integrated Windows authentication is used in IIS because the token is not able to pass between machines.[11] Of course, then within SQL Server the token passed from IIS would need the appropriate database access, including a login and permissions to access the appropriate tables. If possible, this should be done using group accounts rather than user accounts for ease of administration.

The end result would be typical configurations like those shown in Table 7-1, which lists the considerations for the various intersections of IIS and SQL Server security.

---

[10] The topic that describes this is called "Updating the Database of Trusted Certificate Authorities on a Windows-CE based Device" in the SQLCE Books Online.

[11] Delegation of trusted accounts can only occur using Kerberos, which is not supported by SQLCE.

Table 7–1 RDA Security Configurations

	IIS Authentication	
SQL Server Authentication	**Integrated Windows**	**SQL Server**
`Basic`	Can be used with intranet or Internet (when coupled with SSL). Advantage is that a user's credentials can be passed from the device to the database, and the application would need to collect only one set of credentials.	Can be used with Internet (with SSL) or intranet scenarios. Allows the IIS authentication to be decoupled from SQL Server. All users can connect to SQL Server with the same account, enabling connection pooling. Disadvantage is that SQL Server credentials are stored on the device or two sets of credentials must be captured from the user.
`Integrated Windows`	Should be used in intranet scenarios only, where no proxy server is present. Is supported only if IIS is on same computer as SQL Server. Advantage is that no password is passed from the device across the channel.	Should be used for Intranet scenarios, where no proxy server is present. Advantages are that it can be used if the SQL Server does not reside on the same computer as IIS and it decouples the authentication. Disadvantage is that SQL Server credentials must be stored on the device or two sets of credentials must be captured from the user.
`Anonymous`	Should be used with intranet scenarios only, but a service account should be configured explicitly on IIS to connect to SQL Server. Advantage is that connection pooling can be taken advantage of and no credentials are stored on the device or passed over the channel. This option is not considered secure.	Can be used for intranet or Internet scenarios. Advantage is that it requires only one set of credentials to be captured from the user but is considered less secure than basic authentication because it relies only on SQL Server authentication.

TIP: For more information on both RDA and merge-replication security, see the white paper "Security Models for SQL Server 2000 Windows CE Edition 2.0" referenced at the end of this chapter. This white paper leads you through the testing of various security configurations through a sample installation.

For a technique that shows how the relevant properties might be configured, consider the code in Listing 7-1.

Listing 7–1 *Setting Security Properties.* This listing shows a helper class that can be created to set the various properties required by RDA. This class also ensures that only one instance of the `SqlCeRemoteDataAccess` class will be created in the application.

```
Namespace Atomic.SqlCeUtils

    Public NotInheritable Class RDA

        Public Shared ReadOnly Instance As RDA = New RDA
        Private Shared _rda As SqlCeRemoteDataAccess
        Private _sqlCePath As String

        Private Sub New()
            ' Private, so no public instance can be created
        End Sub

        Shared Sub New()
            ' Read the default values from a config file or the registry
            ' and use the appropriate constructor
            _rda = New SqlCeRemoteDataAccess
        End Sub

        Public ReadOnly Property SqlCeRDA() As SqlCeRemoteDataAccess
            Get
                Return _rda
            End Get
        End Property

        Public Property URL() As String
            Get
                Return _rda.InternetUrl
            End Get
```

```
            Set(ByVal Value As String)
                _rda.InternetUrl = Value
            End Set
        End Property

        Public Property Login() As String
            Get
                Return _rda.InternetLogin
            End Get
            Set(ByVal Value As String)
                _rda.InternetLogin = Value
            End Set
        End Property

        Public Property Password() As String
            Get
                Return _rda.InternetPassword
            End Get
            Set(ByVal Value As String)
                _rda.InternetPassword = Value
            End Set
        End Property

        Public Property SqlCePath() As String
            Get
                Return _sqlCePath
            End Get
            Set(ByVal Value As String)
                _sqlCePath = Value
                _rda.LocalConnectionString = _
        "Provider=Microsoft.SQLSERVER.OLEDB.CE.2.0;Data Source=" &
Value
            End Set
        End Property

        Public ReadOnly Property SqlCeConnectionString() As String
            Get
                Return _rda.LocalConnectionString
            End Get
        End Property

       Public Property SqlConnectionString() As String
         Get
             Return _sqlconn
         End Get
```

```
        Set(ByVal Value As String)
            _sqlconn = Value
        End Set
     End Property

    Public Property ProxyServer() As String
        Get
            Return _rda.InternetProxyServer
        End Get
        Set(ByVal Value As String)
            _rda.InternetProxyServer = Value
        End Set
    End Property

    Public Property ProxyLogin() As String
        Get
            Return _rda.InternetProxyLogin
        End Get
        Set(ByVal Value As String)
            _rda.InternetProxyLogin = Value
        End Set
    End Property

    Public Property ProxyPassword() As String
        Get
            Return _rda.InternetProxyPassword
        End Get
        Set(ByVal Value As String)
            _rda.InternetProxyPassword = Value
        End Set
    End Property

    Public Sub SaveSettings()
        ' Save the configuration properties
    End Sub
End Class

End Namespace
```

In Listing 7-1, the `Atomic.SqlCeUtils.RDA` class is a sealed class that uses the Singleton design pattern[12] and composition to expose an instance of

[12] For more information, see the MSDN article referenced at the end of this chapter.

the SqlCeRemoteDataAccess class using the Instance read-only field. Using this pattern all of the code in the application can use the same instance of the class, and it can be configured automatically the first time it is used. This can be accomplished via the registry or a configuration file, which is read in the shared constructor. Because the various security properties are exposed as read/write, they can also be configured through a UI in the application or set programmatically on the instance. You can see that the class also exposes a SaveSettings method to save the configuration settings back to the registry or config file.

Although it is not discussed earlier, the SqlCeRemoteDataAccess class also needs the connection string to the local SQLCE database, here exposed as the SqlCeConnectionString property. A C# client using this class can then programmatically change some of the settings and default the others as follows:

```
RDA.Instance.Password = txtPwd.Text;
RDA.Instance.Login = txtUser.Text;
RDA.Instance.SqlCePath = FileSystem.RuntimeFolder + "\\mylocaldb.sdf";
RDA.Instance.URL = "http://myserver/virtualdirectory/sscesa20.dll";
```

Alert readers will notice that the class contains a read-only property to expose the SqlCeRemoteDataAccess object itself. Although this can be done, a more encapsulated solution would be to add methods to the RDA class to support the Push, Pull, and SubmitSQL functionality of the SqlCeRemoteDataAccess class. In this way, the RDA class acts as a complete wrapper for RDA, and parameters to the various methods such as the connection string to the remote SQL Server can be populated automatically.

Logging

Although RDA will report conflicts on the client if configured to do so, it is also important to log information on the server because it will enable administrators to diagnose problems more easily. To enable logging by the server agent, a particular registry key (HKLM\Software\Microsoft\MSSQLSERVERCE\Transport) and value, pointing to the local path of the virtual directory, must be created on the IIS computer. After IIS is restarted, an XML log file (Sscerepl.log) will be created in the virtual directory. The log file can contain three levels of information specified by the registry value: errors only; errors and warnings; and errors, warnings, and informational messages. The last is recommended only when attempting to troubleshoot a particular problem and contains various timings that are defined in the Books Online.

The log file contains a series of <STATS> elements that are written to the file every few minutes and contain any activity for either RDA or merge replication during that time.

Using RDA

The API for the `SqlCeRemoteDataAccess` object mimics that of the RDA object available to eVB and eVC developers and simply contains the `Pull`, `Push`, and `SubmitSQL` methods.

Pull

The `Pull` method executes any valid Transact-SQL statement on the SQL Server, retrieves the data, creates the associated table in SQLCE from the result set, optionally turns on client tracking in SQLCE, and optionally defines the error table in SQLCE, which will be created during the `Pull` operation and populated if errors are encountered when the data is pushed back to the server.

NOTE: Keep in mind that the Transact-SQL statement needn't return any rows. In this case, the table will still be created in SQLCE and can still be tracked for changes by the SQLCE engine. This strategy is intentionally used when the data originates on the device (such as an order-creation process) and is later saved to the SQL Server using the `Push` method.

Typically, developers would want the SQL statement passed to the `Pull` method to include a `WHERE` clause or to call a stored procedure that includes one. This ensures that only a subset of the server table is retrieved so as not to overload the device and can be used to partition the data horizontally based on the identity of the user, customer, geography, or some other partitioning column. Data can also be partitioned vertically only by including a subset of the columns from the table in the `SELECT` clause.

The SQL statement can also return rows from multiple tables generated through a `JOIN` clause, although doing so does not allow SQLCE to track the changes made to the rows, and so the table cannot be updated using the `Push` method. This is the case with any nonupdatable result set returned from the `Pull` method, and so in these cases, the tracking option (defined as an enumeration) passed to the method must be set to `RdaTrackOption.TrackingOff`. Creating a nonupdatable result set is useful for static lookup data (product and location codes, for example), and the judicious use of JOINs can increase performance by allowing data to be retrieved in one roundtrip to the database server, rather than in two or more.

If the result set is updatable and the table has a primary key defined, tracking can be enabled either with or without other indexes in SQLCE. If indexes are enabled, then the same indexes that exist on the SQL Server table will be created on the SQLCE table, provided that the SELECT clause contains the indexed column or, in the case of a composite index, all of the columns that make up the index. This can obviously improve performance for queries on the device (especially if the Seek method is used as described in Chapter 5) but should be considered in light of how the SQLCE database will be used by the application. Creating unnecessary indexes will serve only to slow performance and consume memory on an already constrained device. In either case, the primary key constraint will be created in SQLCE. With tracking enabled, SQLCE will create two system columns on the table that are used to track the changes made to each row.[13]

Of course, as mentioned previously, the server agent must map the data types returned from SQL Server to those supported by SQLCE. If the result set contains an unsupported data type (such as timestamp), an exception will be thrown by the SqlServerCe Data Provider.

Because of the fairly simple nature of the Pull method, it is relatively easy for developers to implement. For example, the class in Listing 7-1 could be augmented to expose a series of overloaded Pull methods as shown in Listing 7-2.

Listing 7–2 *Calling the Pull Method.* This listing augments the RDA class shown in Listing 7-1 and includes overloaded methods used to call the Pull method.

```
Public Overloads Sub Pull(ByVal sqlCeTable As String, _
 ByVal sqlString As String)
    ' Assume no tracking and no indexes
    _pull(sqlCeTable, sqlString, RdaTrackOption.TrackingOff, Nothing)
End Sub

Public Overloads Sub Pull(ByVal sqlCeTable As String, _
 ByVal sqlString As String, ByVal indexes As Boolean)
    ' Check for using indexes
    If indexes Then
        _pull(sqlCeTable, sqlString, _
```

[13] S_BinaryKey and S_Operation. These columns are protected by the SQLCE engine.

```
            RdaTrackOption.TrackingOffWithIndexes, Nothing)
    Else
        _pull(sqlCeTable, sqlString, _
          RdaTrackOption.TrackingOff, Nothing)
    End If
End Sub

Public Overloads Sub Pull(ByVal sqlCeTable As String, _
 ByVal sqlString As String, ByVal indexes As Boolean, _
 ByVal errorTable As String)

    ' Validate the error table name and check for using indexes
    If errorTable Is Nothing OrElse errorTable.Length = 0 Then
        Throw New ArgumentNullException("Must supply an error table.")
    End If

    If indexes Then
        _pull(sqlCeTable, sqlString, _
          RdaTrackOption.TrackingOnWithIndexes, errorTable)
    Else
        _pull(sqlCeTable, sqlString, _
          RdaTrackOption.TrackingOn, errorTable)
    End If
End Sub

Private Sub _pull(ByVal sqlCeTable As String, _
  ByVal sqlString As String, ByVal tracking As RdaTrackOption, _
  ByVal errorTable As String)

    If sqlCeTable Is Nothing OrElse sqlCeTable.Length = 0 Then
        Throw New ArgumentNullException( _
         "SQLCE table must be specified.")
    End If
    If sqlString Is Nothing OrElse sqlString.Length = 0 Then
        Throw New ArgumentNullException( _
          "A SQL statement must be specified.")
    End If

    ' Add the tracked table to the arraylist
    If (tracking = RdaTrackOption.TrackingOn Or _
      tracking = RdaTrackOption.TrackingOnWithIndexes) Then
        _pulledTables.Add(sqlCeTable)
    End If
```

```
        ' Call the actual pull method
        Try
            _rda.Pull(sqlCeTable, sqlString, _
              Me.SqlConnectionString, tracking, errorTable)
        Catch e As SqlCeException
            ' Log the error
            LogSqlError("RDA.Pull",e)
            If e.HResult = &H80004005 Then
                ' Table already exists
                ' Wrap the exception in a custom object with the message
            Else
                ' Other errors
                ' Wrap the exception in a custom object with the message
            End If
        End Try
    End Sub
```

You'll notice that Listing 7-2 includes three overloaded `Pull` methods. These methods accept the SQLCE table to be created, the SQL statement to execute against the SQL Server, the name of the error table to create if errors are generated when the data is synchronized, and, finally, whether indexes should be created on the device. The three public methods are responsible for validating the arguments and then delegating the actual work to the private `_pull` method and responsible for calling the `Pull` method of the `SqlCeRemoteDataAccess` object and handling any exceptions. As discussed in Chapter 5, if an exception does occur, a best practice is to wrap the exception in a custom exception class derived from `System.ApplicationException` and add custom error messages and other information before throwing the exception back to the caller.

A C# developer using these methods could then write the following code to retrieve a list of teams from a SQL Server and place them in the Teams table in SQLCE:

```
string teamSql = "exec usp_GetTeams";
RDA.Instance.Pull("Teams", teamSql, true);
```

TIP: Of course, an alternative approach is simply to expose the SQLCE table name as an argument to the `Pull` methods and then automatically build the `SELECT` statement, thereby requiring one less argument to the methods. This comes at the cost of flexibility used to specify the columns and selection criteria.

While the `Pull` method is very simple to use, it does have several minor limitations:

- Computed columns are not allowed in the `SELECT` clause, or an exception will be thrown.
- Columns with the attribute `ROWGUIDCOL` are not allowed and must be excluded from the `SELECT` clause.
- Data cannot be pulled from a table having a primary key of type `char`, `nchar`, `varchar`, or `nvarchar`, if they have a length greater than 255 characters.[14]
- SQLCE is not case sensitive, and so, if the SQL Server uses a case-sensitive collation, developers will need to be aware that queries used on the server (or stored procedures invoked to pull rows down to the device) may not return the same data as those on the device.
- The source table on SQL Server cannot be dropped, renamed, have its primary key dropped, or have any columns added, renamed, or dropped if the SQLCE table is tracking changes and will later be synchronized.

TIP: Although the SQL Server table's schema must remain unchanged, the indexes, foreign keys, identity columns, and `DEFAULT` constraints on tracked SQLCE tables can be changed.

Push

After an application creates tracked SQLCE tables and allows the user to manipulate them, changes to those tables can be synchronized with the SQL Server using the `Push` method of the `SqlCeRemoteDataAccess` class. This method simply accepts the SQLCE table to synchronize, the OLE DB connection string for the SQL Server, and a value that specifies whether the changes are to be applied within the context of a single transaction or individual transactions for each row.

Following the previous example, the methods in Listing 7-3 could be added to the RDA class to support synchronizing a table with the SQL Server.

---

[14] This is because those data types will be mapped to the ntext data type, and a primary key cannot be created on `ntext`.

Listing 7–3 *Calling the Push Method.* This listing augments the RDA class shown in Listing 7-1 and includes overloaded methods used to call the `Push` method to synchronize changes on the device with SQL Server.

```
Public Overloads Sub Push(ByVal sqlCeTable As String)
    _push(sqlCeTable, False)
End Sub

Public Overloads Sub Push(ByVal sqlCeTable As String, _
  ByVal batch As Boolean)
    _push(sqlCeTable, batch)
End Sub

Private Sub _push(ByVal sqlCeTable As String, ByVal batch As Boolean)

   Dim bOpt As RdaBatchOption
   If batch Then
        bOpt = RdaBatchOption.BatchingOn
   Else
        bOpt = RdaBatchOption.BatchingOff
   End If

   If sqlCeTable Is Nothing OrElse sqlCeTable.Length = 0 Then
        Throw New ArgumentNullException( _
         "SQLCE table must be specified.")
   End If

   Try
        _rda.Push(sqlCeTable, Me.SqlConnectionString, bOpt)
   Catch e As SqlCeException
        ' Log the error
        LogSqlError("RDA.Push",ex)
       ' Wrap the exception and throw it back to the caller
   End Try
End Sub
```

Once again, the two overloaded public methods expose the SQLCE table that is to be synchronized and an optional argument that determines whether the changes should be made in the context of a single transaction. If the batching option is used, then the server agent starts a transaction on the SQL Server before issuing any `INSERT`, `UPDATE`, or `DELETE` statements.

If any errors occur, the server agent rolls back the entire transaction. When not in batch mode, each row that is synchronized is treated as an independent or implicit transaction. In either case, all rows that cause errors when synchronized with SQL Server are copied to the error table specified in the method, along with the OLE DB error message, the time the error occurred, and the OLE DB error number. This ensures that applications can be aware of all the rows that will have conflicts on the server, even when using batch mode. It is also important to note that because server-level tracking is not supported by RDA, the server agent simply uses optimistic concurrency (it does not lock records when a `Pull` occurs). As a result, RDA defaults to a "last one in wins" scenario, whereby the last device to synchronize data will have its changes persisted to the database. Once again, this is the reason why RDA is best suited for applications whose data is naturally well partitioned.

KEY POINT

It is important to note that the error table is cleaned out before each `Push` method, and so, it will reflect only the errors from the most recent synchronization. In addition, when not in batch mode, rows that cause errors will be deleted from the tracked SQLCE table after being added to the error table. This implies that applications must be designed to read from the error table, allow the user to fix the data, and subsequently merge the data back into the original table to be pushed back to the server at a later time. However, because sophisticated conflict resolution is not supported by RDA, the majority of the errors encountered will likely be caused by connectivity and security problems or rows being deleted on the server.

TIP: Triggers are not supported in SQLCE; however, data modification statements executed against SQL Server during synchronization may cause triggers to fire on the server. For these triggers the `SET NOCOUNT ON` option should be set so that, if the server sends a message indicating no rows were affected, it won't cause the `Push` method to throw an exception.

You could imagine that because the `RDA` class is wrapping the functionality of the `SqlCeRemoteDataAccess` object (`_rda`), the class could be extended to keep track of all the tables that were pulled with tracking enabled and then expose a method to synchronize them all. In fact, the code to track the pulled tables is in the private `_pull` method shown in Listing 7-2, where tracked tables are added to a private `ArrayList`. This `ArrayList` can then be iterated using a `For Each` loop to synchronize all tables in a `PushAll` method as shown in Listing 7-4.

Listing 7–4 *Grouping Synchronization.* This method synchronizes all the tables pulled with tracking in the `RDA` class by iterating an `ArrayList` built when the `Pull` method is called.

```
Public Sub PushAll(ByVal batch As Boolean)
    ' Push all the tracked tables
    Dim bOpt As RdaBatchOption
    Dim tab As String

    If batch Then
        bOpt = RdaBatchOption.BatchingOn
    Else
        bOpt = RdaBatchOption.BatchingOff
    End If

    Dim tab As String
    Dim sqle As New SqlCEUtilException( _
     "An error occurred pushing multiple tables")
    Dim errors As Boolean = False

    For Each tab In _pulledTables
        Try
            _rda.Push(tab, Me.SqlConnectionString, bOpt)
        Catch ex As SqlCeException
            ' Log the error
            LogSqlError("RDA.Push of table " & tab,ex)
            errors = True
            Dim er As New SqlCeUtilError
            er.Message = "Could not push table " & tab & _
             "[" & ex.Message & "]"
            er.Number = 100
            er.Table = tab
            sqle.Errors.Add(er)
        End Try
    Next
    ' Throw the exception if an error occurred
    If errors Then Throw sqle

End Sub
```

The `PushAll` method in Listing 7-2 also ensures that an attempt to synchronize each table is made. If an exception occurs while the table is being processed, it is placed in a custom `SqlCeUtilError` object and added to a collection of errors in a custom exception class (`SqlCeUtilError`). If any of the tables has been unsuccessfully synchronized, the method throws the exception after the loop.

You'll notice that this algorithm could be modified to throw the exception on the first table that causes an error, which would be appropriate if the tables are dependent on each other as in the case of order and order details tables.

Beware of SQL Injection Attacks

Since the Code Red and Nimbda attacks of 2001, Microsoft has become much more focused on security, as evidenced by their "Trustworthy Computing" initiative. This has led to products that are secure by default and, therefore, reduce the attack surface for applications built by organizations like yours; however, individual developers also need to be aware of secure coding practices.

One such example is protection against SQL injection attacks. These are attacks whereby a hacker attempts to encode valid, but often destructive, SQL in UI elements that are then unwittingly executed by an application to reveal or destroy confidential data. For example, a typical approach is for a hacker to include a Transact-SQL end of literal string character ('), followed by some destructive SQL, then to append a line comment (–) to invalidate any other Transact-SQL that may come later. These attacks can work on both SQL Server and SQLCE.

To avoid SQL injection attacks, developers should code their applications with the cardinal rule, don't trust user input, in mind. This means that anywhere an application embeds user input in a SQL statement, that input should be checked for validity. For example, to avoid the particular situation just discussed, the developer could replace all single quotes the user entered with two single quotes. Another more generic approach is to use the classes of the `System.Text.RegularExpressions` namespace to validate the user input using regular expressions so that special characters can be disallowed.

Although most Compact Framework applications are not as inherently prone to such attacks as public Web sites because of their lower profile (the applications are typically used by a smaller user base, only when the application is installed on the device), Compact Framework developers using RDA should still take note. This is because the `Pull` and `SubmitSql` methods simply accept SQL statements and do not use command and parameter objects, as do the SqlClient and SqlServerCe Data Providers (discussed in Chapters 4 and 5), which inherently protect against such attacks.

For more information on secure coding practices, see Howard and LeBlanc's book referenced in the "Related Reading" section at the end of this chapter.

SubmitSql

The final method exposed by the `SqlCeRemoteDataAccess` class and the simplest to use is the `SubmitSql` method. When invoked, the server agent simply executes the Transact-SQL statement passed as an argument to the method against SQL Server. The only caveat is that the method cannot return a result set. The method is quite useful for executing stored procedures in SQL Server to perform data modifications and data cleanup tasks invoked after a successful synchronization. The `SubmitSql` method of the RDA class developed in this chapter is shown in Listing 7-5.

Listing 7–5 *Submitting Transact-SQL.* This method submits a SQL statement using the `SubmitSql` method of the `SqlCeRemoteDataAccess` class.

```
Public Sub SubmitSql(ByVal sqlString As String)
    ' Validate the argument
    If sqlString Is Nothing OrElse sqlString.Length = 0 Then
        Throw New ArgumentNullException("Must supply a SQL statement")
    End If

    Try
        _rda.SubmitSql(sqlString, Me.SqlConnectionString)
    Catch ex As SqlCeException
        ' Log the error
        LogSqlError("RDA.SubmitSql with SQL of " & sqlString,ex)
       ' Wrap and throw the exception
    End Try
End Sub
```

Merge Replication

In the final section of this chapter, we'll explore the second of the two synchronization mechanisms available through SQLCE, merge replication. Like RDA, the client agent provides an API that applications use to submit replication requests (add and drop a subscription, reinitialize a subscription, and synchronize) over HTTP to the server agent. As with RDA, Compact Framework developers access the API through the `SqlCeReplication` class in the SqlServerCe .NET Data Provider.

The subscription includes information about an existing publication created on a SQL Server (referred to as the publisher). The publication consists of individual articles or tables that have been enabled for replication.[15] In short, and as shown in Figure 7-3, when an application creates a subscription using the `AddSubcription` method, an initial snapshot of the tables (schema and data) is downloaded to the device from a distributor[16] by the server agent, and the tables are created in SQLCE by the client agent. The SQLCE database can then be updated by a Compact Framework application, while not connected as in RDA, thereby making it ideal for occasionally connected scenarios. At some point later, the application reconnects to the network and calls the `Synchronize` method of the `SqlCeReplication` class. This method instructs the client agent to extract all the modified rows in the subscription tables and send them to the server agent. On the IIS server, the server agent then creates an input file with all of the `INSERT`, `UPDATE`, and `DELETE` requests from the device. A process on the IIS server called the SQL Server Reconciler[17] then loads a special replication provider for SQLCE,[18] which reads the input file and tells the Reconciler about the changes. The Reconciler then makes the appropriate changes on the publisher and detects and resolves conflicts created by other subscribers updating the same row. This happens according to the rules specified on the publisher.[19]

TIP: It should be noted that SQLCE can act only as a subscriber and not as either a publisher or a distributor.

The Reconciler then tells the SQLCE replication provider about the changes that must be made to the subscription database. The provider writes these changes to an output file, which is picked up by the server

---

[15] Technically, SQL Server publications can also include other objects, such as stored procedures, user-defined functions, and views; however, SQLCE supports only tables.

[16] An instance of SQL Server that is used to store the snapshots, history, and statistics. This can be the same server as the one on which the data resides.

[17] Also referred to as the merge agent.

[18] This is a pluggable architecture that allows other replication providers, such as the one for SQL Server subscribers, to process synchronizations as well.

[19] There are standard conflict resolvers available on SQL Server that can be configured for the publication to handle simple cases. It is also possible to write custom resolvers to handle more complex logic. For more information, see the SQL Server 2000 Books Online.

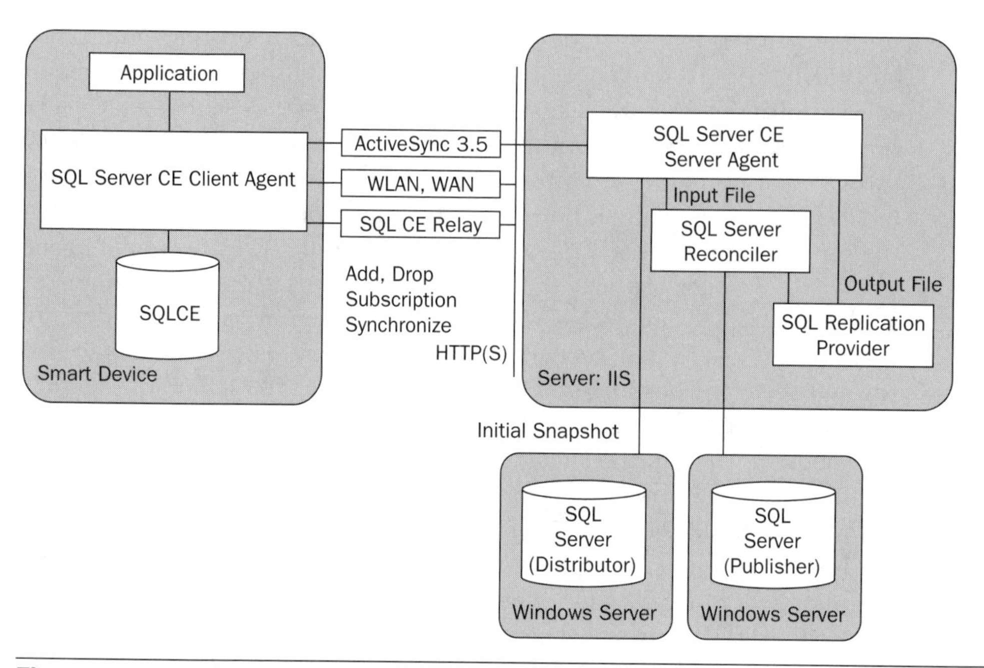

Figure 7–3 *Merge Replication Architecture*. This diagram shows the architecture of merge replication and highlights the components involved.

agent and sent to the client agent on the device. The client agent then applies these changes to the subscription tables.

KEY POINT

Obviously, this process is referred to as merge replication because changes made on the device are merged with changes made on other devices and with those made directly on the publisher (for example, through a Web application updating the database). As a result, merge replication is ideal when a database receives continuous updates or is updated both by the client and the server or by multiple clients.

Features and Scenarios

Merge replication includes the following advantages that make it attractive for certain types of solutions:

- *Efficient resource usage:* Merge replication takes advantage of the connectivity architecture discussed previously. In addition to the 8:1 compression that conserves bandwidth and ability to work when the underlying transport is not particularly reliable, this architecture is used to initialize subscriptions and synchronize changes from the server to the device. This allows the client agent to process blocks of data and discard them, rather than requiring storage memory for large snapshots and sets of changes.

- *Horizontal and vertical partitioning:* Although RDA allows partitioning by formulating queries passed to the `Pull` method, the various articles within the publications can include row and column filters. This allows administration of the filters to be controlled on the server (and controlled by an administrator), rather than on the client, making administration and development simpler. In addition, merge replication supports *dynamic filtering* (discussed later in the chapter), whereby the articles contain dynamic logic used to filter based on information provided by the subscriber. In this way the data to be downloaded to the device can be customized by user, group, or geography.
- *Multiple tables:* Unlike RDA, which supports only pulling and pushing data from one table at a time, individual articles, each representing a table, can be placed in a publication. This allows multiple tables and their data to be created on the device when a subscription is created, thereby making development simpler; however, this flexibility comes with the additional administrative overhead of creating publications on the server.
- *Server-based conflict resolution:* As described previously, individual articles can be configured with a supplied[20] or custom conflict resolver. These resolvers, which execute on the distributor, encapsulate the rules for determining which subscriber's data should be favored when more than one subscriber updates the same row. This allows the rules that govern conflicts to be centralized on a server rather than existing within the application; however, because of the algorithm required to merge changes between multiple subscribers, SQLCE must track more information for each row. Whereas RDA requires only one column (4 bytes) per row, replication requires three columns (48 bytes) of overhead per row.
- *Bidirectional data flow:* Perhaps the most significant feature, however, is that replication supports a bidirectional data flow. In other words, while RDA allows tables to be pulled from SQL Server and changes made on the device to be synchronized with the server, it does not support flowing changes made on SQL Server onto the device without repulling the entire table. Merge replication supports this bidirectional flow by default, although it can be limited to synchronize

---

[20] There are actually 10 resolvers that ship with SQL Server 2000. See the SQL Server 2000 Books Online for more information about what each does.

changes originating only on the device. This feature makes merge replication much more efficient for applications that require incremental updates from the server.

The combination of these features means that merge replication is most appropriate in the following scenarios:

- *Modifying shared data:* Because merge replication includes server-based conflict detection and resolution that can be customized to any level of complexity, it is well suited for situations where the data is not highly partitioned, so that multiple subscribers may need to update the same row. An example might be the need to view and update customer information that overlaps sales territories.
- *Working with dependent data:* Because multiple tables (articles) can be grouped into a single publication and made available to the device, merge replication works well when the tables have some dependency between them. For example, an application that requires customer and order history information can subscribe to a publication that includes both as articles, rather than having to make separate calls using RDA.
- *Occasionally connected applications:* Because merge replication does not support any real-time or connected way of directly querying the SQL Server, it is inherently suited for situations where the data can be updated only through the local SQLCE database on the device. This occasionally connected architecture is appropriate for Compact Framework applications that run on devices where there is unreliable or no wireless connectivity and synchronization occurs only sporadically, for example, through a docking station.
- *Working with changing data:* Even if the application does not perform updates, merge replication is efficient for pushing out changes made on the server. Because the client agent downloads only incremental changes, there is not the overhead associated with repulling an entire table, as would be the case with RDA. For example, the Multiple Listing Service (MLS) application discussed in the following sidebar is a good example.

Configuration

In order to use merge replication, a publication must first be created on the SQL Server. The simplest way to accomplish this is to use the Create Publication Wizard in SQL Server Enterprise Manager. This wizard leads an

SQL Server CE and Real Estate

One of the success stories involving SQL Server CE 1.1 that has gained significant publicity since 2002 has been the Pocket PC MLS application from Offutt Systems, called Pocket InnoVia. This application allows agents to download MLS listings and contact information, including photos of the property, directly to their Pocket PCs for offline viewing.

This system uses merge replication to download the listings and associated data based on a geographic location to the Pocket PC. New listings are then added to the device incrementally when the application performs a synchronization. This system also stores its final data in Oracle and uses Data Transformation Services (DTS) in SQL Server 2000 to refresh the publication database periodically.

Perhaps the most significant aspect of this case study, however, is that because SQLCE includes the built-in replication infrastructure, the development costs were kept under $100,000, while the low administration costs allow Offutt Systems to sell the application for $129 and to charge $50 for a per-user, per-year subscription.

Although the system was originally developed with eVB and SQLCE 1.1, it is being updated to use the Compact Framework and SQLCE 2.0 to take advantage of better query performance, better server performance, and additional features such as parameterized queries. It will also take advantage of the Pocket PC Phone Edition and Short Message Service (SMS), in conjunction with SQL Notification Services, to provide instant notification of new listings and one-touch calling of agents.

For more information on this case study, see the "Related Reading" section at the end of the chapter.

administrator step by step through the process of selecting the database to publish, the appropriate publication type (merge publication in this case), the types of subscribers expected (shown in Figure 7-4), the articles to publish, and other properties, including filters. If this is the first publication created on the server, it also allows an administrator to identify which server will act as the distributor.

By selecting the Devices running SQL Server CE option in the dialog shown in Figure 7-4, the wizard automatically allows anonymous subscribers for the publication and will create the snapshot data using character mode format. Both of these options are required in order for devices to create subscriptions, and they can also be configured once the publication has been created by right-clicking on it and selecting Properties in the context menu.

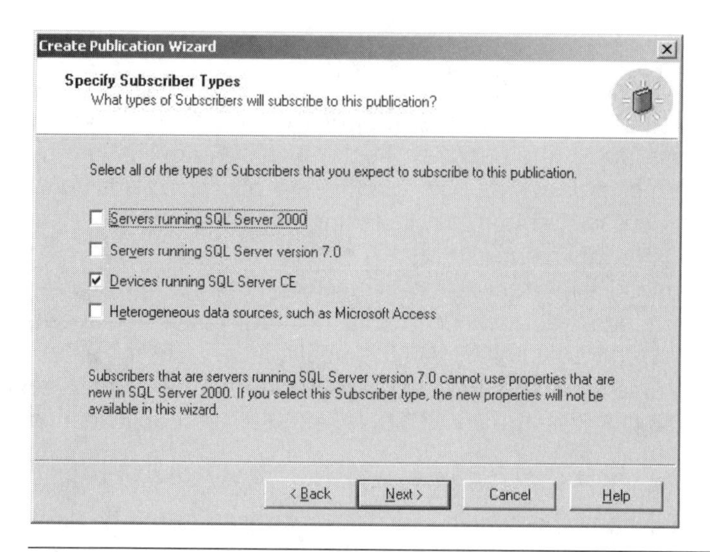

Figure 7–4 *Selecting Subscriber Types.* This dialog appears in the Create Publication Wizard and is used by the wizard to enable anonymous subscribers and character mode snapshots, if SQL Server CE subscribers is checked.

In addition, although merge publications support either column- or row-level tracking, only row-level tracking is recognized by the SQLCE replication provider. As a result, if all subscribers will be devices running SQLCE, an administrator should configure the publication accordingly for efficiency. It is also important to configure an explicit snapshot folder on the distributor. This is recommended because the snapshot folder will default to an administrative share (`C$`) accessible only by administrators. Because the server agent running on IIS needs to access this share, it makes sense to create an explicit nonadministrative share and then configure the distributor to use this share to make the snapshot available.[21]

As mentioned previously, one of the features of merge replication is that row and column filters can be configured on the server and that the row filters can be dynamic in nature. To configure these, the Create Publication Wizard leads the administrator through the process of creating the filters. For example, if row filtering is chosen, the dialog shown in Figure 7-5 allows the administrator to choose further that dynamic filters will be enabled.

The wizard then presents the dialog shown in Figure 7-6 to create the dynamic filter. The important point to note about the filter is that it relies on

[21] See the topic "Configuring an Explicit Snapshot Folder" in the SQLCE Books Online.

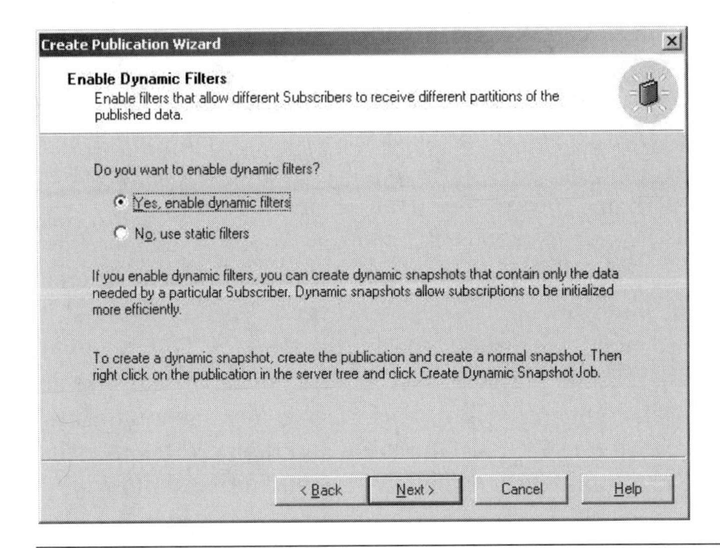

Figure 7–5 *Selecting Dynamic Filters.* This wizard screen allows an administrator to create dynamic filters on a publication.

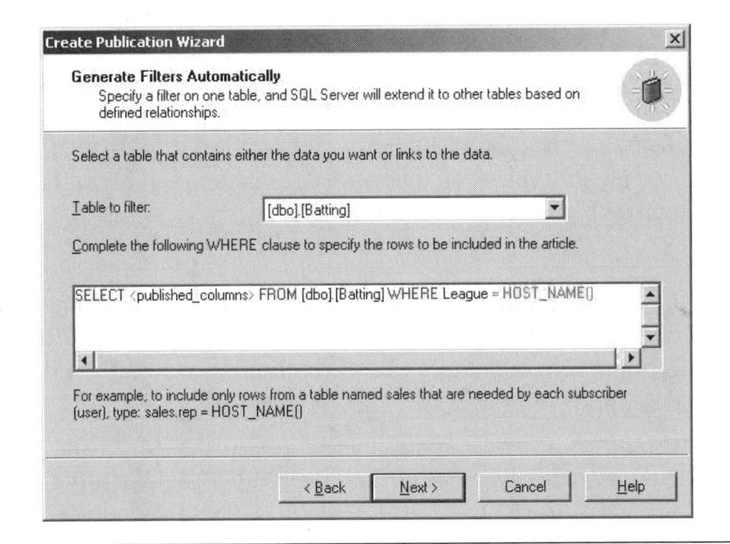

Figure 7–6 *Creating the Dynamic Filter.* This dialog allows the administrator to create the dynamic filter based on the return value of the `HOST_NAME` T-SQL function.

the `HOST_NAME()` Transact-SQL function in the `WHERE` clause to evaluate which rows are created in the initial snapshot and subsequently updated on the subscriber.[22] This value is populated by the `HostName` property of the `SqlCeReplication` class on the device, thereby allowing dynamic views of the article. Typical values with which to populate the `HostName` property include user name, region, department, and product line.

When dynamic filters have been created, the default behavior is for dynamic snapshots to be generated as subscriptions are created by the Compact Framework application; however, to speed the initial synchronization time, dynamic snapshots can be precreated. This is accomplished by right-clicking on the publication in SQL Enterprise Manager and selecting Create Dynamic Snapshot Job. This wizard allows an administrator to specify a value for the `HOST_NAME` function and then creates the snapshot agent to create, and optionally to schedule, the initial snapshot.

NOTE: Although not mentioned previously, one of the other useful features of merge replication is support for auto-ranged identity columns. This feature allows tables with system-assigned primary keys (identity columns) to be published and instructs the server to manage the range of values that can be inserted at the publisher and individual subscribers. A tab in the Article Properties dialog allows the administrator to configure a range size for the publisher and each subscriber along with a threshold at which to assign new ranges. This feature ensures that subscribers do not attempt to insert different rows with the same identity value. An alternate approach would be to use natural keys or the `uniqueidentifier` data type.

Although configuring merge replication for Compact Framework applications is fairly straightforward, it does entail using a subset of the features available to other subscribers. Some of the limitations that need to be considered follow:

- *No alternate synchronization partners:* Once a subscription has been created on the device, the Compact Framework application must always synchronize with the original publisher. In other words, alternate synchronization partners are not supported for SQLCE.

[22] The `suser_name` function is also supported by dynamic filters, but since SQLCE subscribers are always anonymous subscribers, that function will not return a unique value for the user.

- *Restricted use of identity columns:* Because SQLCE supports only identity columns of type integer, subscribing to an article that includes identity columns of any other data type will fail.
- *No computed columns:* SQLCE does not support creating computed columns, and so, subscribing to tables that contain computed columns will automatically create a vertical partition that excludes those columns.
- *No case sensitivity:* As in RDA, it is important to remember that SQLCE does not support case-sensitivity, and so, primary keys and unique indexes based on case-sensitive data in a publication cannot be created on SQLCE.
- *No null values in unique constraints:* Because SQLCE does not allow `NULL` values in a column with a `UNIQUE` constraint, while SQL Server does, subscriptions cannot be created on SQLCE if the publication contains `NULL` values in a column with a `UNIQUE` constraint.
- *`NOT FOR REPLICATION` ignored:* Because SQLCE does not support this option, any constraints that are marked `NOT FOR REPLICATION` will still be created on the SQLCE subscriber.

Security and Connectivity

Although many of the security settings discussed in the RDA section of this chapter also apply to merge replication, creating a publication also entails configuring several other security-related settings:

- *Database access:* Because the server agent on IIS invokes the SQL Server Reconciler, the identity of the process on IIS must have permissions on the database in SQL Server. For example, if anonymous authentication is selected in IIS, coupled with Windows authentication in SQL Server, then the `IUSR_computername` account on the IIS server must be given permission to access the database. Likewise, if basic or integrated authentication is chosen in IIS and Windows authentication in SQL Server, then the account specified in the `InternetLogin` property of the `SqlCeReplication` class must be provided access. Finally, if SQL Server authentication is used, then the account specified by the `PublisherLogin` property of the `SqlCeReplication` class must be given access.
- *PAL:* Each publication on the SQL Server is also associated with a list of logins that SQL Server uses to restrict access to the publication during synchronization. The account configured in the previous bullet point must also be added to the PAL for publication by right-clicking

the publication in the Enterprise Manager and clicking the Publication Access List tab. Optionally, an administrator can also enable the Check Permissions option in the Article Properties dialog to force SQL Server to ensure that the SQL Server Reconciler has permission to perform `INSERT`, `UPDATE`, and `DELETE` statements on the publication database.

In addition to the security settings mentioned in this list, the `SqlCeReplication` class supports configuring connectivity and security through a series of properties. For example, the `PublisherSecurityMode` property can be used to specify the authentication mode of the publisher. If the property is set to `SecurityType.DBAuthentication`, then the `PublisherLogin` and `PublisherPassword` properties must be specified so that the SQL Server Reconciler on IIS can use these to login to the publisher. Additionally, the `PublisherNetwork` and `PublisherAddress` properties can be used to configure which network protocol and address the Reconciler can use to connect to the publisher if the default protocol configured on IIS should not be used. The only required properties, however, are `Publisher` and `PublisherDatabase`, which specify the name of the publishing server and the name of the publication database, respectively.

In the same way the `SqlCeReplication` class supports a series of distributor properties, none of which is required if the publisher and distributor are on the same computer, but which are used by the Reconciler to connect to the distributor.

As you can imagine, these properties are best configured using the system registry or a configuration file on the database to make them easier to update. As a result, a class that wraps the initialization and creation of `SqlCeReplication` objects using a simple version of the prototype design pattern[23] would be useful. A skeleton of such a class is shown in Listing 7-6.

Listing 7–6 *Wrapping `SqlCeReplication`.* This class acts as a factory for creating instances of `SqlCeReplication` and configuring them with the appropriate properties.

```
public sealed class ReplicationFactory
{

    private ReplicationFactory {}
```

[23] Documented in *Design Patterns* and referenced in the "Related Reading" section at the end of this chapter.

```
static ReplicationFactory
{
        /*
         Read the configuration properties from the
         registry or config file.

         These would include at a minimum:
         Publisher, PublisherDatabase, PublisherSecurityMode
         SubscriberConnectionString, Subscriber
         InternetLogin, InternetURL

         And the following if necessary:
         PublisherLogin, PublisherPassword
         PublisherAddress, PublisherNetwork
         InternetProxy properties if necessary
         Distributor properties
         LoginTimeout, QueryTimeout
        */
}

public static SqlCeReplication CreateCeReplication()
{
    SqlCeReplication rep = new SqlCeReplication();
    _initClass(ref rep);
    return rep;
}

public static SqlCeReplication CreateCeReplication(
 string publication)
{
    SqlCeReplication rep = new SqlCeReplication();
    _initClass(ref rep);
    rep.Publication = publication;
    return rep;
}

public static SqlCeReplication CreateCeReplication(
  string publication, ExchangeType exchangeType)
{
    SqlCeReplication rep = new SqlCeReplication();
    _initClass(ref rep);
    rep.Publication = publication;
    rep.ExchangeType = exchangeType
    return rep;
}
```

```
    private static void _initClass(ref SqlCeReplication rep)
    {
            // set the properties of the rep object based on the values
            // read in the constructor
    }
}
```

As you can see, the `ReplicationFactory` class reads in the default properties from a configuration file or the registry in the static constructor. These could optionally be exposed as static properties that can be set programmatically. It then exposes overloaded `CreateCeReplication` methods to create the replication objects populated with the default properties, as well as, optionally, the name of the publication and the exchange type (bidirectional or upload only). Obviously, there are a number of permutations of both the configuration properties to set and the arguments to expose in methods such as `CreateCeReplication`. The class designer's choice of which to set and expose depends on such factors as the security configuration and whether the application supports multiple subscriptions. In this case the class assumes that different publications or exchange types may be configured at runtime.

A client can then use this class as shown in the following snippet, which will return a fully functional `SqlCeReplication` object that is ready to add the subscription and begin synchronization.

```
Dim myRep as SqlCeReplication
myRep = ReplicationFactory.CreateCeReplication( _
 "CustomerData", ExchangeType.BiDirectional)
```

Topologies

From an architectural perspective, there are several topologies that an organization can implement with merge replication. The four most common are briefly described here and shown in Figure 7-7.

1. *Single server:* With this option, the publisher, distributor, and server agent with IIS all reside on a single machine. While this configuration certainly puts more load on the loan server and is somewhat less secure, it is easy to configure and can be suitable for applications with few users or when the publisher is used simply to collect data that is later used to update another database via DTS. In this configuration the server can reside behind a firewall.

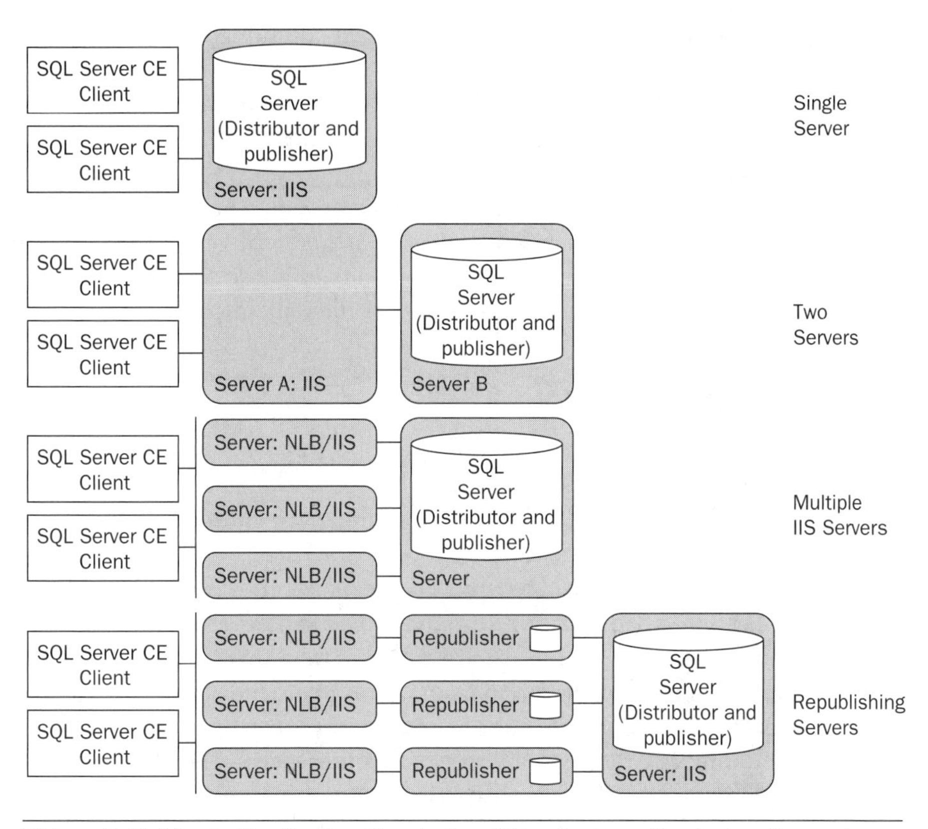

Figure 7–7 *Merge Replication Topologies*. This diagram illustrates the various topologies that can be used with merge replication and SQLCE.

2. *Two servers:* In this configuration, the server agent and IIS reside on a server separate from the publisher and distributor (both running on the same SQL Server). With this option, the load on the server is reduced, and the IIS server can reside in a DMZ behind a firewall with another firewall protecting the publisher or distributor. This option enhances security and is used when there are a limited number of concurrent synchronizations.
3. *Multiple IIS servers:* If the system needs to support many concurrent synchronizations, the two-server solution can be scaled out by implementing a Web farm of IIS servers using load-balancing hardware or software such as the NLB of Windows 2000 Server. NLB can be used because it allows the initial client agent request to be redirected to one of the servers in the cluster and then ensures that the conversation with the server agent on the selected server is preserved (referred to as client affinity, as mentioned previously).

4. *Republishing servers:* Finally, if additional load balancing is required, a series of republishing servers can be configured. These servers act as subscribers to the original publisher and then republish the data through their own merge publications. This offloads processing from the original publisher; however, this option introduces additional configuration concerns because the SQLCE does not support alternate synchronization partners. In other words, each SQLCE subscriber would need to be configured statically with the name of the publisher, making the load balancing less dynamic.

Using Replication

Once the publication is configured on the server and the connectivity and security properties configured in the Compact Framework application, actually creating the subscription and performing the synchronization are relatively straightforward.

The `SqlCeReplication` class exposes the `AddSubscription`, `ReinitializeSubscription`, and `Synchronize` methods to create the subscription, reapply the publication's snapshot, and invoke replication, respectively. For example, in the previous code snippet, the `ReplicationFactory` was used to create a `SqlCeReplication` object with all of its properties set for the `CustomerData` publication. To then create the subscription and apply the initial snapshot, the following code can be executed:

```
Dim myRep as SqlCeReplication
myRep = ReplicationFactory.CreateCeReplication( _
 "CustomerData", ExchangeType.BiDirectional)

Try
    myRep.AddSubscription(AddOption.ExistingDatabase)
    myRep.Validate = ValidateType.NoValidation
    myRep.Synchronize()
Catch ex As SqlCeException
    ' Wrap the exception and report the error
End Try
```

TIP: Alternatively, the `ReplicationFactory` class could expose additional overloaded `CreateSubscription` methods that create and initialize the `SqlCeReplication` object in addition to calling the `AddSubscription` and `Synchronize` methods.

You'll notice that the `AddSubscription` method can either use the existing database specified in the `SubscriberConnectionString` property or create a new one. Additionally, as shown in the snippet, the `Validate` property specifies whether the synchronization process is validated by comparing the row count of the published data on the server with that on the subscriber.[24] This option causes locking to occur on the publisher as row counts are checked, and so, typically this option would be used at some interval to ensure that the subscriber is still in sync with the publisher. If the validation fails, the code can call the `ReinitializeSubscription` method to reapply the snapshot to SQLCE.

TIP: `ReinitializeSubscription` must also be executed when using dynamic filters if the `HostName` property is changed.

After a synchronization occurs, the `SqlCeReplication` object will populate the `PublisherConflicts`, `PublisherChanges`, and `SubscriberChanges` properties with row counts that detail the changes that were made. These properties can then be displayed through the UI of the Compact Framework application.

It's important to keep in mind that conflicts detected by the SQL Server Reconciler will be handled by the resolvers configured on the publisher. While there are tracking tables that can be accessed on the server that allow conflicts and their resolutions to be viewed, unlike RDA, SQLCE does not track those conflicts on the device.

What's Ahead

This chapter has focused on the two built-in synchronization technologies that are a part of SQLCE and that are accessible to Compact Framework applications: RDA and merge replication. This chapter therefore ends Part II of this book, which has covered the four essential architectural concepts of accessing local data, accessing remote data, caching data locally, and synchronization. The remaining chapters in Part III address additional programming

---

[24] SQL Server also supports validating by calculating a checksum, but SQLCE only supports using the row count value. As a result, the `ValidateType` enumeration only supports the `RowCountOnly` value.

considerations, starting with globalization and localization of Compact Framework applications.

Related Reading

"Security Models and Scenarios for SQL Server 2000 Windows CE Edition," at http://msdn.microsoft.com/library/default.asp?url=/library/en-us/dnsqlce/html/sqlce_secmodelscen20.asp.

"Troubleshooting SQL Server CE Connectivity Issues," at www.microsoft.com/sql/techinfo/administration/2000/CEconnectivitywp.asp.

"SQL Server 2000 Windows CE Edition and the .NET Compact Framework," at http://msdn.microsoft.com/library/en-us/dnsqlce/html/sqlwince.asp.

"Exploring the Singleton Design Pattern," at http://msdn.microsoft.com/library/default.asp?url=/library/en-us/dnbda/html/singletondespatt.asp.

Howard, Michael, and David LeBlanc. *Writing Secure Code*. Microsoft Press, 2002. ISBN 0-7356-1588-8. Chapter 12 discusses SQL injection attacks and other typical Web-based attacks.

"SQL Server CE Case Studies," at www.microsoft.com/sql/ce/productinfo/casestudies.asp. Many of these case studies involve using RDA, merge replication, or both.

"Pocket InnoVia SQL Server CE Case Study," at www.microsoft.com/resources/casestudies/CaseStudy.asp?CaseStudyID=13404.

Gamma, Erich, et al. *Design Patterns*. Addison-Wesley, 1995. ISBN 0-201-63361-2. Chapter 3 discusses the various creational patterns and what they are used for.

PART III

Additional Programming Considerations

CHAPTER 8

Localization

Executive Summary

In our ever more connected world, many organizations require that the software they write be used in different regions and cultures. In order to cut down on the costs of localization, provide the ability to localize an application rapidly, and reduce maintenance by using a single code base across localized versions, software architects and developers need to design and build software with globalization (the process of designing an application to be "world ready") and localization (the process of adapting the software to a particular culture or region) in mind. Fortunately, the Compact Framework takes much of the work out of globalization and localization by directly supporting culture-specific settings on the device and by being able to localize resources in satellite assemblies.

Both the Compact Framework and its desktop cousin identify cultures using an industry-standard hierarchical nomenclature in the form *language code-country/region code*. The Compact Framework then abstracts various culture-specific settings for formatting dates and times, strings, and numeric data, including currency, into `System.Globalization.CultureInfo` objects. The default culture set on the device uses the regional settings and is accessible using the static `CultureInfo.CurrentCulture` property. Unlike the desktop Framework, however, the `CurrentCulture` cannot be set on a thread-by-thread basis, and so developers must create and store specific `CultureInfo` objects in order to use a culture other than the default programmatically.

Within an application, developers then often need to display localized versions of `DateTime`, `String`, and numeric data to users. Fortunately, the `ToString` methods of the built-in data types and others, those that sort strings, for example, take the current culture into account. Many of these methods are also overloaded to accept a `CultureInfo` object that they then use to format the data appropriately. Some considerations have to be made by developers when localizing data, especially when dealing with currency, which cannot for obvious reasons be directly converted.

In order to separate the UI from the code in a world-ready application, the Compact Framework relies on resource files and, fortunately, does much of the work by automatically creating resource-specific assemblies (complete with a fallback system), referred to as satellite assemblies, which are culture-specific and can be deployed with the application. By simply creating culture-specific resource files in VS .NET 2003, assemblies containing the localized resources will be created when the project is built. Developers can then read these localized resources from the satellite assemblies using the `GetString` and `GetObject` methods of the `ResourceManager` class in the `System.Resources` namespace.

The Compact Framework also provides localized exception text assemblies that can be used, and developers can test their world-ready applications with localized versions of the emulator images.

The Need for Globalization and Localization

In the ever more competitive global marketplace, it is important for many organizations to be able to design software that can be used not only in the United States, but in a variety of other cultures as well. In many cases, organizations that will use the Compact Framework are based in or have offices in Canada, Mexico, South America, Europe, and elsewhere in the world. In these cases, developing an application that is globalized, or world ready, has three primary benefits:

1. *Rapid adaptation:* If the application as been designed to be world ready, then its adaptation, or localization, to a new culture will be relatively straightforward. This means that it can be more quickly deployed and, in the end, affect the bottom line in a positive way.
2. *Development cost savings:* Not only can localized applications be deployed to new cultures and regions more quickly, but the cost of localization in terms of developer effort and expense for each new culture is greatly reduced. This is obviously the case because designing an application with globalization in mind up front, although slightly more expensive during initial development, means that the application itself should not have to be redesigned or redeveloped to support additional cultures. As a result, the incremental cost of developing localized resources for additional cultures does

not entail nearly the costs of redesigning and redeveloping the application to support localization.

3. *Single code base:* Creating an application that is globalized also has the advantage of allowing an organization to support a single code base. This is obviously more efficient as bug fixes, changes, and enhancements need to be made to only one set of code. It also reduces the errors because attempting to keep multiple sets of code in sync always creates additional errors. A side benefit here as well is that localizing strings as required for a globalized application allows the application to be more easily modified to, for example, change the strings used in dialog boxes.

Fortunately for organizations that want to develop world-ready applications, the Compact Framework supports both globalization (the process of designing an application that is world ready by separating application functionality from localizable resources, such as UI elements and regional data) and localization (the process of creating and applying regional data in a globalized application).

As we'll discuss in this chapter, the Compact Framework takes much of the work out of globalization and localization by directly supporting culture-specific settings on the device and by being able to localize resources in satellite assemblies.

Guidelines for Globalization and Localization

Before discussing the features of the Compact Framework that support globalization and localization, it is important to keep a few basic guidelines[1] in mind, especially if you or your team has little experience in this area.

- *Avoid political and cultural hot buttons.* As you can imagine, when designing an application to be world ready, you and your team should avoid any issues that might be culturally or politically sensitive. This can include the use of colloquialisms, slang, images that are ethnocentric, and even maps that include contested regions. Any of these items can cause your application to be rejected by users in a particular region, even though it may function as designed.

---

[1] Other guidelines can be found at the Microsoft site listed in the "Related Reading" section at the end of the chapter.

- *Isolate the UI*. As we'll discuss later in the chapter, developers and architects should strive to isolate all UI elements, meaning the messages for prompts and dialog boxes, for example, from the rest of the program code. This can be done through resource files and satellite assemblies.
- *Be aware of string and screen sizing*. It is important when designing the UI of a world-ready application to be aware of the sizes of messages in various languages. If a prompt in a `Label` control in English requires a size of 130 pixels, that same prompt may require 260 pixels in another language. Leaving enough space on the window or relying on dynamic sizing techniques will allow your application's UI to display correctly. Of course, testing the application with the most common or target cultures is also required.
- *Avoid using text in bitmaps and icons*. Be careful to avoid embedded text in icons and bitmaps. Often, this text is in English, and an entirely different image would be required in a different culture. Although this can be done, a simpler technique would be to create icons and images without text and, if text is necessary, to isolate the associated text in a resource file.
- *Beware of word order*. Because word order varies among languages, you or your developers should avoid string concatenation that builds coherent messages or that uses insertion parameters in a format string. Both of these can lead to garbled text in a variety of languages.

Globalization and Localization Support

The Compact Framework includes support for globalizing applications and then localizing them for specific cultures and regions. Like many things in the Compact Framework, the features to accomplish this are a subset of those found in the desktop Framework. Specifically, when designing and implementing a world-ready application, there are three primary features of the Compact Framework that must be understood: cultures, localized data, and the use of satellite assemblies.

Understanding Cultures

In order to identify various cultures and regions, the Compact Framework and its desktop cousin both use the hierarchical nomenclature laid out in RFC 1766. This document specifies that each language code is specified by

Table 8–1 Culture Names

Culture Name	Language–Country/Region	Locale/Culture Identifier
En	English	0x0009
en-US	English–United States	0x0409
en-GB	English–United Kingdom	0x0809
en-CA	English-Canada	0x1009
Fr	French	0x00C
fr-FR	French-France	0x040C
fr-CA	French-Canada	0x0C0C
De	German	0x0007
de-AT	German-Austria	0x0C07
de-DE	German-Germany	0x0407

a two letter code[2] that acts as the parent of a two- or three-letter country or region code.[3] Thus, "en" is the code for English and "US" is the code for the United States. Together the culture (or locale) name "en-US" is used to refer to English, United States. Table 8-1 lists some of the western European culture names. These names are hierarchical in nature in the form *language/country* or *region*.

As you can see from Table 8-1, the bilevel nature of the culture names allows for those cases where only a language group, such as English, is identified and others where a country, such as Canada, where both English and French are spoken and which contains regional differences in those languages, is identified. As a result, the culture names without country or region specified are referred to as neutral, whereas the presence of country or region code makes them specific.[4] In addition to specifying the language, a culture name also identifies the formats for time, currency, and numeric data, as well as the calendar used.

You'll also notice that Table 8-1 includes the locale/culture identifier (LCID) that uniquely identifies the culture to the Compact Framework and that is also used by Windows operating systems such as Windows XP and Windows CE. For example, the Pocket PC 2002 supports over 90 cultures.

---

[2] Derived from ISO 639-1

[3] Defined in either ISO 3166 or ISO 639-2

[4] There are some exceptions. For example, "zh-CHS" refers to simplified Chinese, which is a neutral culture. A complete list of cultures supported by the desktop Framework can be found in the Visual Studio .NET 2003 online help.

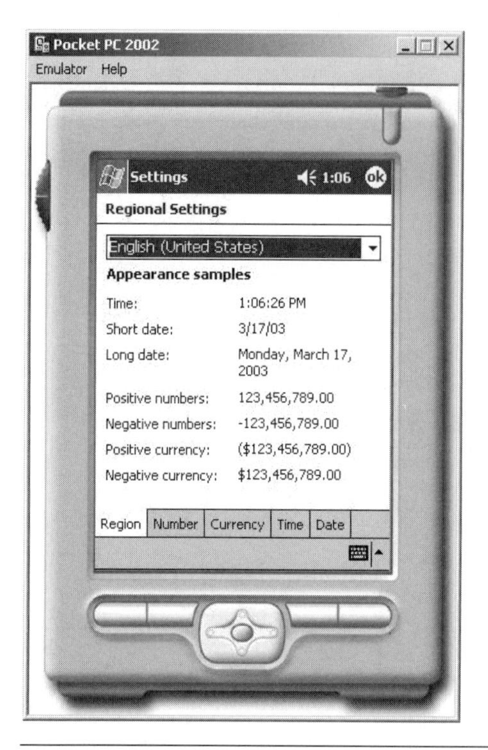

Figure 8–1 *Setting the Culture.* The culture to be used on the device may be set from the Regional Settings application. Note that the culture identifies not only the language, but also the number, currency, time, and date formats that can be customized by the user.

The culture to use on a device such as the Pocket PC 2002 can be chosen from the Regional Settings application as shown in Figure 8-1.

In addition to neutral and specific cultures, the Compact Framework also defines the invariant culture, which is identified by an empty string ("") and LCID 0x007F, is associated with the English language, and is used to represent culturally nonsensitive data. Conceptually, the invariant culture is the parent of all the neutral cultures.

Retrieving and Setting the Culture

In the Compact Framework, as in the desktop Framework, cultural information is represented by the `CultureInfo` class in the `System.Globalization` namespace. This class contains static (shared in VB) methods to retrieve the

CurrentCulture and the CurrentUICulture. Both properties are set based on the regional settings defined on the device as shown in Figure 8-1. Although both of these properties are set identically, the CurrentCultureUI property is present because operating systems that have a multilanguage user interface (MUI), such as special versions of Windows 2000 and Windows XP, can set this property to differ from CurrentCulture.

To determine the current culture programmatically, a developer can access the methods and properties of the CultureInfo object exposed by the CurrentCulture property. For example, to print the English name of the current culture, its date-time format, and whether it is a neutral culture, the following code could be added to an SDP:

```
MessageBox.Show(CultureInfo.CurrentCulture.EnglishName)
MessageBox.Show( _
    CultureInfo.CurrentCulture.DateTimeFormat.FullDateTimePattern)
MessageBox.Show(CultureInfo.CurrentCulture.IsNeutralCulture.ToString())
```

However, in the Compact Framework the CurrentCulture property is read-only, unlike in the desktop Framework, where it can be modified and where each thread can execute using a different culture, as exposed through the CurrentCulture property of the System.Threading.Thread class. This restriction is in place because smart devices such as the Pocket PC are single-user devices and, therefore, don't require multiple threads of execution that service different clients to use different cultures.

However, a developer can create instances of the CultureInfo class and then store the instances in class variables for use in the application (for example, if the device is shared by multiple users[5]). This technique can be used to allow the user to switch the culture without changing the regional settings, or when the default culture for a specific user is stored in a database. For example, to create a business class that can be customized to use different cultures dynamically, the constructor of the class can be overloaded to accept a culture name, LCID, or a CultureInfo object as shown in Listing 8-1.

---

[5] This may especially be the case when the application is used in regions such as Belgium, where multiple official languages are the norm.

Listing 8–1 *Using Specific Cultures.* This listing shows a business class that accepts a specific culture in its constructor and stores that culture in a private class variable.

```
Public Class SomeBusinessClass

    Private _culture As CultureInfo

    Public Sub New()
        ' Do other work
        _culture = CultureInfo.CurrentCulture
    End Sub

    Public Sub New(ByVal culture As String)
        ' Do other work
        Try
            _culture = New CultureInfo(culture, True)
        Catch e As PlatformNotSupportedException
            ' Default to current culture
            _culture = CultureInfo.CurrentCulture
        End Try
    End Sub

    Public ReadOnly Property Culture() As String
        Get
            Return _culture.EnglishName
        End Get
    End Property

    ' Other overloaded constructors can accept LCID or a CultureInfo
End Class
```

In this case the default constructor simply picks up the current culture from the `CurrentCulture` property, while the parameterized constructor accepts the culture name. Note that the `CultureInfo` class will throw a `PlatformNotSupportedException` if the culture name is not valid for the Compact Framework or for the underlying operating system. Finally, you can see that when the `CultureInfo` object is created, the second argument to this version of the constructor specifies whether to use the user-customized values for the culture. By specifying True, as in this case, if a user has customized the time format in the regional settings on the device, for example, those settings will be used. By specifying False, the Compact Framework will use its default values.

As a result, the class could be instantiated as follows to use the en-CA (English-Canada) culture:

```
Dim biz As New SomeBusinessClass("en-CA")
```

NOTE: This technique of implementing a constructor in a class that sets the culture to be used can also be employed to great effect in an abstract or base class. In this way, all derived classes (for example, those in the data-services or business-services layers) can benefit from this functionality.

Localizing Data

One of the important steps developers must take when globalizing an application is to check how data (dates and times, strings, currency, and other numbers) is displayed to the user. Fortunately, the Compact Framework does much of the work based on the current culture, although there are a few items that architects and developers need to be aware of.

Handling Dates and Times

Displaying dates and times (as represented by the `System.DateTime` data type) appropriately when different culture settings are used is handled automatically by the Compact Framework, when the `ToString`, `ToShortDateString`, `ToShortTimeString`, `ToLongDateString`, or `ToLongTimeString` methods are called on a `DateTime` object.[6] For example, when the culture is set to en-US, the `ToShortDateString` method returns 4/8/2003; however, when en-CA is used, the date is displayed as 08/04/2003.[7]

Developers can also format dates and times using a specific culture by using the overloaded `ToString` method. One of the overloads accepts a format string, while another accepts an object that implements the `IFormatProvider` interface. The `CultureInfo` class exposes a `DateTimeFormat` property (actually a `DateTimeFormatInfo` object) that implements the `IFormatProvider` interface and then exposes string properties, such as `ShortDatePattern` and `FullDateTimePattern`, that return the various

---

[6] When the `ToString` method is called, the resulting string contains the concatenation of the short date and short time strings.

[7] The four-digit year is returned here even though in Windows CE, two-digit years are the default for short dates. This is because the Compact Framework has its own set of default values that override those of the underlying operating system.

patterns for dates and times for the specific culture. By passing one of the pattern properties, or simply a string that contains a custom pattern, to the ToString method, the DateTime object is formatted. By passing the DateTimeFormatInfo object to ToString, the short date is used. Additionally, the ToString method accepts one-character identifiers that map to the various patterns; for example, "d" maps to the ShortDatePattern, while "t" maps to ShortTimePattern. The pattern and the DateTime-FormatInfo objects can be used together in ToString to produce a formatted DateTime value for a specific culture:

```
string s = startDate.ToString("d", CultureInfo.InvariantCulture);
```

In the reverse, the DateTime class exposes Parse and ParseExact methods that can be used to convert a formatted string to a DateTime value. By passing the format string or the DateTimeFormatInfo object to the methods along with the string, the DateTime will be created.

For example, the class shown in Listing 8-1 could include a StartDate-String property like the following:

```
Private _startDate As DateTime

Public Property StartDateString() As String
    Get
       Return _
          _startDate.ToString(_culture.DateTimeFormat.ShortDatePattern)
    End Get
    Set(ByVal Value As String)
       _startDate = DateTime.Parse(Value, _culture.DateTimeFormat)
    End Set
End Property
```

The property shown in this snippet accepts a string and uses the DateTime.Parse method to interpret the string using the ShortDatePattern, which is the default when the DateTimeFormatInfo object is passed to the method. If the string is invalid for the format, a FormatException will be thrown. The property then returns the DateTime formatted as a short date using the ToString method.

KEY POINT

A second issue that arises when handling date and time values in a world-ready application is that different users operate in different time zones. For example, you can easily foresee a scenario where a DateTime value is input by a user on a device with its clock set to use U.S. central standard time, and the device synchronizes with a SQL Server using merge replication. Later, another device operating in U.S. eastern standard time pulls down the record using merge replication and displays it to the user. In this case, the user in the eastern time zone will not be viewing a local time, and, unless the application has some other means of tracking where the date was recorded, it will be difficult to reconcile the problem.

To handle this situation the **DateTime** value should be stored in SQL Server CE as a universal **DateTime** value. And, as a best practice, the **DateTime** value should be translated to a universal format anytime the value is persisted. Fortunately, the Compact Framework makes this easy by exposing the **ToUniversalTime** and **ToLocalTime** methods on the **DateTime** structure. These methods convert local **DateTime** values to **DateTime** values that use coordinated universal time (UTC) or Greenwich mean time (GMT) and back again. This universal value can then be stored in SQL Server CE's **shortdatetime** and **datetime** data types and, when read back, converted to the local time for display.

The **DateTimeFormatInfo** object also contains a **UniversalSortableDateTimePattern** property that can be used with the **ToString** method to create a string that can be persisted and still sorted appropriately. Of course, the patterns associated with the **InvariantCulture** can also be used with **ToString** to produce culture-independent strings.

As an example of storing **DateTime** values in a universal format, the **StartDateString** property of the class shown in Listing 8-1 could be amended to store the start date internally as a universal date time, so that when it was saved to a database or file, it would already be in the universal format.

```
' Stored as universal datetime
Private _startDate As DateTime

Public Property StartDateString() As String
    Get
        Return _startDate.ToLocalTime.ToString( _
          _culture.DateTimeFormat.ShortDatePattern)
    End Get
    Set(ByVal Value As String)
        _startDate = DateTime.Parse(Value, _
          _culture.DateTimeFormat).ToUniversalTime
    End Set
End Property
```

Now, you'll notice that the **ToUniversalTime** method is called when the string is parsed, in order to save the **DateTime** value as universal, and the ToLocalTime method is called to convert it back to the time zone configured on the device for display.

The final issue to consider in terms of localizing date and time information is the use of calendars. Simply put, each culture as represented by a CultureInfo object exposes a Calendar property that contains an object derived from Calendar. The Compact Framework supports only calendars based on the Gregorian calendar (the one used in the United States and the Western world), which includes the GregorianCalendar, JapaneseCalendar, ThaiBuddhistCalendar, KoreanCalendar, and TaiwanCalendar. The desktop Framework additionally supports non-Gregorian calendars, such as the Julian, Hebrew, and Hirij calendars. For Gregorian-based calendars, typically only the year number and era differ. Each calendar object supports methods and properties to return the year, era, and other information. For example, to return the current year using the ThaiBuddhistCalendar, the following code could be written to return 2546 for the Gregorian year 2003.

```
ThaiBuddhistCalendar c = new ThaiBuddhistCalendar();
MessageBox.Show(c.GetYear(DateTime.Now));
```

As discussed previously, the best practice for dealing with varying calendars is to accept input based on the calendar associated with the current culture and then persist the value in the universal format. When converting back again using the ToLocalTime method, the correct calendar will be used to display the year and era.

Handling Strings

As with date and time values, the Compact Framework automatically takes into consideration the current culture when dealing with strings in terms of fonts, comparisons, and character casing. For example, when the String.Compare, Array.Sort, TextInfo.ToUpper, and TextInfo.ToLower methods are called, the current locale is consulted. The information about string comparisons is stored in the CompareInfo property of the CultureInfo object, which exposes methods that are used by methods such as String.Compare.

NOTE: As you might have guessed, if the application needs to perform string comparisons that are culture independent (for example, those that are used for security to activate or deactivate features in the application), the InvariantCulture.CompareInfo methods should be used so that the behavior will be consistent, regardless of the current culture.

However, because string comparisons and, therefore, sort orders differ between cultures,[8] a world-ready application should perform sorting in the application code rather than in a database like SQLCE. For example, rather than return a sorted `SqlCeDataReader` using a SQL statement (described in Chapter 5) with a `WHERE` clause that is then used to bind to a `ComboBox`, a developer would want to read the data into an `ArrayList` and then sort the data based on the culture before binding the `ArrayList` to the `ComboBox`, as follows:

```
ArrayList ar = new ArrayList();

// Read the data into an ArrayList
while (dr.Read())
{
    ar.Add(dr["ProductName"]);
}

// Sort using the current culture automatically
ar.Sort();

// Bind
ComboBox1.DataSource = ar;
```

This is the best practice because using a `WHERE` clause relies on the collation defined for the SQLCE database. Because SQLCE's collation may cause the sort order to differ from that of the culture and because SQLCE does not support case-sensitive sorting in any case, the resulting sort order will not be the same as for the culture.

In addition to relying on the current culture, developers can make culture-specific comparisons using the overloaded versions of the `String.Compare`, `TextInfo.ToUpper`, and `TextInfo.ToLower` methods. For example, if a class contained a private variable that referenced the culture, the following code could be written to compare two strings:

```
Dim intResult As Integer = _
  String.Compare(productA, productB, False, _culture)
```

In this case, the string comparison is done based on the `CultureInfo` object referenced in the `_culture` variable; it returns –1 if the first string

[8] For example, the `umlaut` character (as in Ä), sorts differently in different cultures.

should be sorted before the second, 1 if the second should be sorted before the first, and 0 if the strings are equal. The second argument in the call to `Compare` indicates that the comparison should be case sensitive.

This concept can be put to more general use when developers create custom classes used to represent data. For example, the simple `Player` class used to hold information about baseball players, shown in Listing 8-2, implements the `IComparable` interface and, hence, the `CompareTo` method. As with the class shown in Listing 8-1, it accepts a specific culture in one of its overloaded constructors and stores it in a class variable. When instances of the `Player` class are placed into an `Array` or `ArrayList` and sorted, the `CompareTo` method is invoked, and in this case uses the `String.Compare` method to specify that the collection should be sorted on the player's name in a case-insensitive fashion, using the appropriate culture settings.

Listing 8–2 *Sorting Custom Classes.* This listing shows a custom class that implements the `IComparable` interface to do custom sorting based on the culture the class uses.

```
Public Class Player
    Implements IComparable

    Public Name As String
    Public AB, R, H, D, T, HR, RBI, BB, SB As Integer

    Private _culture As CultureInfo

    Public Sub New(ByVal name As String)
        Me.Name = name
        _culture = CultureInfo.CurrentCulture
    End Sub

    Public Sub New(ByVal name As String, ByVal culture As CultureInfo)
        Me.Name = name
        _culture = culture
    End Sub

    ' Used when the Sort method of the collection is called
    Public Function CompareTo(ByVal o As Object) As Integer _
     Implements IComparable.CompareTo
        Dim p As Player = CType(o, Player)
```

```
        Return String.Compare(Me.Name, p.Name, True, _culture)
    End Function

End Class
```

TIP: More general techniques for implementing sorting for custom classes using strongly typed collection classes and custom sort classes can be found in the article listed in the "Related Reading" section at the end of this chapter.

Handling Currency and Other Numbers

The support for formatting and displaying numeric data for specific cultures in the Compact Framework is analogous to that for dates and times. The `CultureInfo` object exposes a `NumberFormat` property that references an instance of the `NumberFormatInfo` class. This class contains all the properties that together define how numeric data should be displayed.

As with the `DateTime` structure, the numeric data types shown in Table 2-2, including `System.Int32`, `System.Double`, `System.Decimal`, and `System.Single`, expose an overloaded `ToString` method that by default uses the current culture, but they can also accept a format string, an `IFormatProvider`, or both. The overloaded `ToString` method can then be used by developers to format numeric data for specific cultures. For example, the following code snippet uses the French (France) culture to display a `System.Double` and would in this case display the value 0,302:

```
CultureInfo c = new CultureInfo("fr-FR");
double avg = 0.302;

MessagBox.Show(avg.ToString(c));
```

This works because the `CultureInfo` class implements the `IFormatProvider` interface. Alternatively, hard-coded format strings can be passed to the `ToString` methods where "g" (or "G") represents a general number, "d" decimal format, "e" floating point, "c" currency, "x" hexadecimal, and "n" number format. And so, for example, the line of code below produces the same result as that above:

```
MessageBox.Show(avg.ToString("g", c));
```

Unlike the pattern properties for the DateTimeFormatInfo class, however, the pattern properties for NumberFormatInfo are integers, rather than strings, and map to hard-coded ways of representing numbers. For example, the NumberNegativePattern property can be set to numbers from 0 to 4 and determines where the negation symbol is placed relative to the number. These properties can be modified on the fly and will therefore affect how the numbers are displayed.[9]

A developer might then use the ToString method to provide a read-only property on a class that returns the culture appropriate format for a numeric value. For example, in the Player class shown in Listing 8-2, the developer might create a read-only Avg property that calculates and returns the player's batting average like so:

```
Public ReadOnly Property Avg() As String
    Get
        Return (H / AB).ToString("g3", _culture)
    End Get
End Property
```

In this case, the format string "g3" is used to show only the first three decimal places, as is standard with batting averages.

As with DateTime values, the reverse can also be done using the Parse method exposed by the various numeric data types. This overloaded method (in addition to accepting the string value to be converted) can accept one or more values from the System.Globalization.NumberStyles enumeration, which specifies what is permitted in the string value and an IFormatProvider object. This method can be used, for example, to convert currency values captured as strings to System.Decimal values:

```
Dim salary As Decimal
salary = Decimal.Parse(txtSal.Text, NumberStyles.Currency)
```

In this case the current culture will be used, but, optionally, a CultureInfo object could be passed as the third argument.

On its face, handling currency values in world-ready applications is no different than handling other numeric data. The "c" format string allows for formatting a numeric value as currency, and the currency properties of the

[9] Consult the VS .NET 2003 online help for the complete list of these mappings.

NumberFormatInfo class, such as CurrencyPositivePattern and CurrencySymbol, contain the formatting information; however, unlike other numbers that should be automatically formatted for the specific culture, currency values, once assigned, are usually always associated with the particular currency. This is the case, of course, because "$1,567.55" and "£1,567.55" are not equivalent, and their true values relative to each other fluctuate dynamically. As a result, unless the developer has access to currency exchange rates (which on a smart device will rarely if ever be the case, although a judicious use of Web Services would solve this problem), currency values should be displayed in the currency in which they were entered. This entails persisting the currency in a database such as SQLCE or simply forcing all currency values to be entered in a standard way, for example, using U.S. dollars.

KEY POINT

The other issue that developers need to be aware of when using currency values is the use of the euro. The Compact Framework defaults the respective currency symbols to "€" for nations that have adopted the euro, including France, Germany, Belgium, Spain, Greece, Italy, and Austria, among others. The confusing aspect is that the various operating systems on which the Compact Framework runs may default the currency symbols to their national currencies, including "F" for France, "DM" for Germany, and "pta" for Spain. For example, the Pocket PC 2000 devices will default to the national currencies, while Pocket PC 2002 will default to the euro. To insure that the euro is used when dealing with specific cultures, the second argument to the CultureInfo constructor can be set to False to indicate that the regional settings on the device are to be ignored, like so:

```
Dim c As CultureInfo = New CultureInfo("fr-FR", False)
```

Using Resources and Satellite Assemblies

The key to separating the UI from the code in a world-ready application is to use resource files. Fortunately, the Compact Framework, like its desktop cousin, does much of the work for developers by automatically creating resource-specific assemblies (complete with a fallback system), referred to as satellite assemblies, which are culture-specific and can be deployed with the application. The Compact Framework also provides the ResourceManager class in the System.Resources namespace to read these resources from the satellite assemblies.

Creating Resource Files

As many developers are aware, resources are simply string or encoded binary values that can be collected in a resource file associated with a project. In Visual Studio .NET 2003, these resource files have a .resx extension and are stored as XML. Developers can create them simply by choosing to add a new item to a project and choosing Assembly Resource File. The resulting resource file can then be edited through a built-in resource editor, as shown in Figure 8-2.

Although the built-in resource editor can be used to add string resources, to add binary resources developers may instead want to use the ResEditor sample application that ships with VS .NET 2003. This application, along with several other localization samples, ships in the \Program Files\Visual Studio .NET 2003\SDK\v1.1\samples\tutorials directory. This

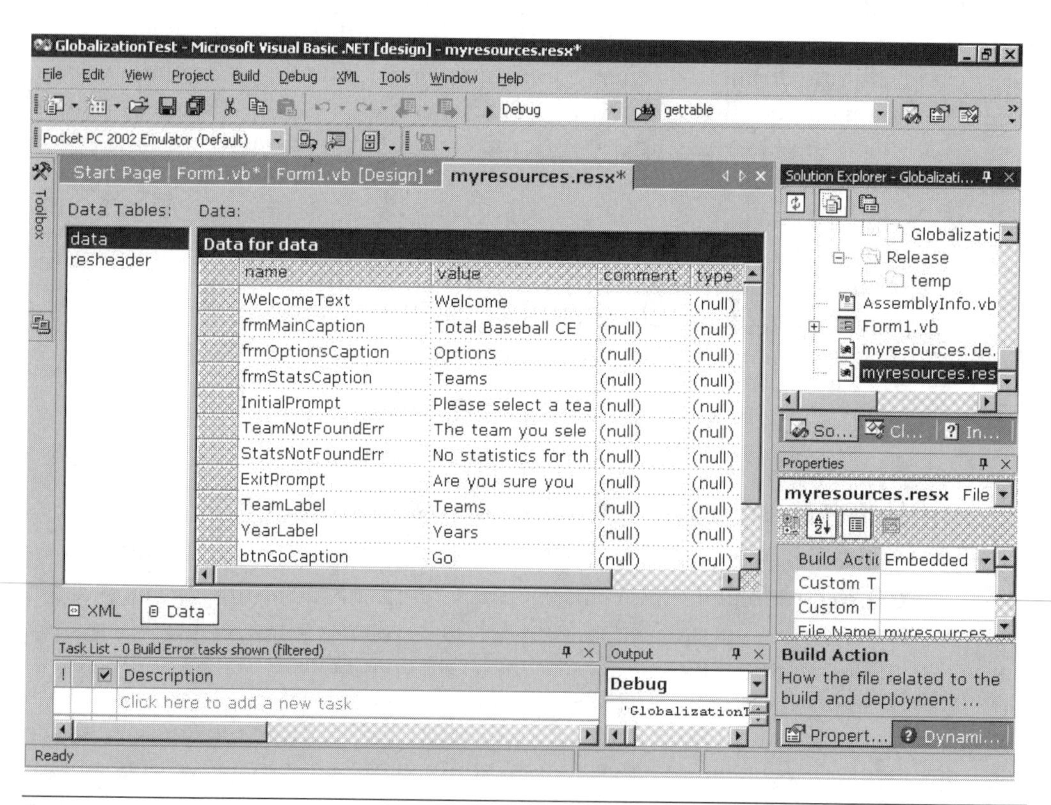

Figure 8–2 *Editing a Resource File.* This screen shot shows the VS .NET IDE's resource editor. It can be used to add simple string resources.

more robust editor works in a stand-alone fashion and can be used to add bitmaps, icons, strings, and image lists to a resource file.[10]

After the resource file has been created and the resources added to the file, VS .NET will automatically compile the resources into a .resource file and embed them in the application's assembly as a type when it is built, because its Build Action property, shown in the properties window, will default to Embedded Resource. By default, the resources in this file are the fallback resources. In other words, if culture-specific resources are not found, those in the fallback .resx file compiled into the assembly will be used.

To create culture-specific resources to perform the task of localization, a developer would then create a resource file for each culture that contains the same resource identifiers as those in the fallback resource file, name the file with the same name as the fallback .resx file, and append the culture name. For example, if the fallback resource file was named MyResources.resx, the resource file containing French resources (a neutral culture) would be called MyResources.fr.resx, and those for the specific culture French (Canada) would be MyResources.fr-CA.resx. As mentioned previously, these resource files should contain culture-specific strings, such as error messages, prompts, labels, titles, and button captions, as well as any bitmaps, icons, and other images that vary.

By simply creating these culture-specific resource files, at build time VS .NET will create a subdirectory with the culture name under the appropriate directory (depending on whether the Debug or Release build is chosen) for each culture-specific resource file. In this directory it places a satellite assembly that contains only the resources in the culture-specific resource file. In the example used above, VS .NET would create an fr-FR directory into which it would place the satellite assembly. As a result, the satellite assembly can be deployed with the application using the same relative pathing structure.

Retrieving Resources

Creating the culture-specific or localized resource files is only half the battle, however. The resource must then be extracted from the resource files at runtime by application code that is globalized. This can be accomplished using the `ResourceManager` class.

[10] For example, there is another sample called ResXGen which generates a resource file from a single image.

To use the ResourceManager class, a developer needs to create an instance of the class and pass overloaded constructor information to it about the resources to retrieve. For example, the following code could be used to create a ResourceManager instance needed to retrieve resources from the MyResources.resx file:

```
Dim res As New ResourceManager("MyAppName.MyResources", _
  [Assembly].GetExecutingAssembly)
```

In this case the constructor accepts the fully qualified type of the resources, which defaults to the root (or default) namespace of the project (here, MyAppName) and the name of the resource file (MyResources). The second argument is a Type object that identifies the assembly in which to find the resources, in this case simply the currently executing assembly.

TIP: One of the overloaded constructors of the ResourceManager class allows developers to use their own custom resource file format. This can be done by passing a reference to the type of a class derived from ResourceSet that developers can use to read and write their own formats.

Once a reference to a ResourceManager object has been created, its GetString and GetObject methods can be used to extract the case-sensitive string and binary resources from the file. For example, if there is a string resource called "frmMainCaption" in the resource file, the following line of code would be used to extract it and set the form's caption:

```
frmMain.Text = res.GetString("frmMainCaption")
```

In an analogous way, binary resources such as images can be read using the GetObject method:

```
pbLogo.Image = CType(res.GetObject("Logo"), Image)
```

In this case the Image property of a PictureBox control named pbLogo is populated. Because GetObject returns the type System.Object, it must be cast to the Image data type.

As with the methods used to localize strings and numbers, both the GetString and GetObject methods are overloaded to support passing in a specific culture using a CultureInfo object, as in the following snippet, where _culture holds a reference to a CultureInfo object.

```
pbLogo.Image = CType(res.GetObject("Logo", _culture), Image)
```

From a design perspective, since the **ResourceManager** instance may need to be available to multiple forms or classes, a good technique would be to expose the instance as a Singleton object using the Singleton design pattern discussed in Chapter 7. A typical Singleton class might look like that shown in Listing 8-3.

Listing 8–3 *Exposing a ResourceManager.* This class implements a Singleton that exposes the ResourceManager object in a shared property.

```
Public NotInheritable Class MyResources

    Private Shared _res As ResourceManager

    Private Sub New()
    End Sub

    Shared Sub New()
        ' Create the instance of the resource manager
        _res = New ResourceManager("MyAppName.MyResources", _
         [Assembly].GetExecutingAssembly)
    End Sub

    Public Shared ReadOnly Property Instance() As ResourceManager
        Get
            Return _res
        End Get
    End Property

End Class
```

In Listing 8-3 the **MyResources** class exposes a single shared (static in C#) property called **Instance** that is initialized in the shared constructor and that exposes the **ResourceManager** reference. In this way, the remainder of the application need not create **ResourceManager** objects and can instead use the Singleton as shown here:

```
frmMain.Text = MyResources.Instance.GetString("frmMainCaption")
```

Another technique for architects to consider is to encapsulate as much as possible of both the `ResourceManager` class and calls to `GetString` and `GetObject`. For example, a UI helper class might expose read-only properties whose `Get` blocks use the appropriate `ResourceManager` object and make calls to `GetString` to retrieve the localized resources.

Other Localization Issues

In addition to localizing data within an application and UI elements, there are two other issues that architects and developers need to consider: exception handling and testing.

Localized Exception Strings

As mentioned in Chapter 2, in order to save space on the device, the text messages associated with exceptions in the Compact Framework are abstracted into a separate assembly called System.SR.dll. This 90K assembly can be referenced during development and, hence, deployed to the device to debug an application more easily.

The Compact Framework also provides nine additional localized versions[11] of the exception messages, each packaged in a .cab file and installed on the developer's computer in the \Program Files\Microsoft Visual Studio .NET 2003\CompactFramework\SDK\v1.0.500\Windows CE\Diagnostics directory. These .cab files, named, for example, System_SR_fr.cab for the French version, also contain the English (U.S.) versions of the messages. As a result, these .cab files can be deployed during either development or production, especially if the exception messages are to be displayed to the end user.

Testing World-Ready Applications

A second issue that should be addressed is the testing of world-ready applications. By default, the emulators that ship with SDP include English images of both the Pocket PC 2002 and Windows CE .NET. As mentioned in Chapter 2, these emulators can be configured through VS .NET and customized in appearance through the use of skins.

[11] These include French, German, Italian, Japanese, Korean, Portuguese (Brazil), Spanish, and two forms of Chinese.

In addition to the two default images, additional localized images can be obtained through the Pocket PC 2002 and Windows CE .NET SDKs.[12] Starting the emulators manually using these images as described in Chapter 2 allows developers to test their applications more fully before they are used in different cultures and reduces the amount of hardware required for testing. For example, the Pocket PC 2002 SDK includes nine localized images encompassing the same languages as those the Compact Framework provides for exception messages, each of which is 15MB to 24MB in size and can be downloaded separately.

Although we certainly recommend testing in this fashion, the best-case scenario, if at all possible, is to test the application on localized devices where possible.

What's Ahead

This chapter has focused on the Compact Framework features and techniques that architects and developers can use to globalize and localize their applications to make them world ready. These include localizing data, such as date and time values, strings, and currency, as well as creating localized resource files. In the next two chapters we'll look at two other considerations for building applications on the Compact Framework: security and deployment options.

Related Reading

"Microsoft Globalization Guidelines," at www.microsoft.com/globaldev/default.mspx.

Fox, Dan. "Take Advantage of Strongly Typed Collection Classes." Builder.com (October 25, 2002), at www.builder.com.

"Pocket PC 2002 SDK," at www.microsoft.com/mobile/developer/downloads/ppcsdk2002.asp.

[12] The URL to download the Pocket PC 2002 SDK can be found in the "Related Reading" section at the end of this chapter.

CHAPTER 9

Securing Compact Framework Solutions

Executive Summary

In recent times, and with good reason, as attacks such as Code Red, Nimbda, and Slammer have shown, security has been pushed to the forefront of the minds of architects and developers. And just as Microsoft has rededicated itself to security by launching its Trustworthy Computing initiative, your organization should be sure to design and implement its applications using the secure coding practices and principles reviewed in this chapter. Even though Compact Framework applications may at first appear immune to such attacks, their connection to back-end systems and their portability introduce a variety of security risks that must be mitigated. As a result, you need to consider security at the device, application, and communication layers.

Securing a device can involve implementing power-on authentication, antivirus protection, and locking down the device. These options are not strictly a part of the Compact Framework but may be implemented by the operating system, for example, in the case of the Pocket PC 2002, which implements power-on authentication, or third-party vendors that provide and extend these facilities through biometrics, smartcards, and other forms of identification. Securing the device protects primarily against unauthorized access to applications should the device be lost, stolen, or infected with a virus.

Securing the application involves implementing authentication, providing data security through encryption, and securing user input. In many cases the facilities used to communicate with back-end systems allow the passing of credentials. The SQLCE Connectivity used with RDA and merge replication and the credentials object used when making HTTP requests are prime examples. In other cases, the Compact Framework application must prompt the user for credentials and use techniques such as SOAP headers to transmit the credentials to an XML Web Service. Protecting the data an application uses through encryption is also important and can be done automatically when using SQLCE or by using third-party products and code libraries. Securing user input involves taking heed of the maxim "Don't trust user input" by preventing SQL injection attacks and making judicious use of regular expressions.

Finally, securing the communications channel can involve using SSL, VPNs, and even custom SOAP extensions when communicating with XML Web Services. The Compact Framework transparently supports both SSL and VPNs through the support offered by host operating systems such as Windows CE. While SSL provides encryption of the communication channel only when using HTTP, VPNs provide both authentication and encryption using protocols such as Point-to-Point Tunneling Protocol (PPTP) and Internet Protocol Security (IPSec). Additional support for securing communications can be obtained by using third-party software. Although many devices connect to corporate networks via Wired Equivalent Privacy (WEP), the WEP algorithm is now considered nonsecure and should be augmented with other techniques such as VPNs.

Security Issues and Principles

Just as the increased internetworking among organizations has driven the need for globalization and localization discussed in the previous chapter, the increased internetworking among computer systems has increased the need to take security seriously. One need look only at the Code Red and Nimbda attacks of recent years to understand the havoc wreaked by malicious people and the software they write. And for every public incident, a myriad of internal security breaches are discovered.

KEY POINT

And just as Microsoft has rededicated itself to security by launching its Trustworthy Computing initiative,[1] your organization should also design and implement its applications using secure coding practices. This is the case because even applications written with the Compact Framework are vul-

[1] Microsoft put its entire development staff through several months of intensive security training and, as a result, postponed shipping key products including Windows Server 2003 and Yukon (the next release of SQL Server) following a memo by Bill Gates. See the security information on Microsoft's site in the "Related Reading" section at the end of the chapter for more information.

nerable to exploitation due both to their portability and to their need to connect with back-end systems to exchange data. To give just one example, Gartner Group estimated that a quarter of a million handheld devices would be left behind or lost at U.S. airports in 2001.

However, before discussing the specifics of securing Compact Framework applications, a brief review of some core security principals or maxims is in order either to refresh your memory or to orient your mind to issues you'll need to face when developing Compact Framework or, indeed, any other kind of software.

- *Apply defense in depth.* This principle asserts the often-repeated phrase that security is only as strong as its weakest link. Because most software is designed in layers and applications incorporate encryption, firewalls, Web servers, and more, architects and developers need to take all of these layers into consideration.
- *Reduce the attack surface.* This principle is a restatement of the obvious: It's easier to hit a barn than it is a tool shed. By disabling features of software that can be attacked, the size of target area for malicious users is reduced.
- *Don't trust user input.* This is perhaps the principle that developers can be most proactive about. As discussed in the sidebar "Beware of SQL Injection Attacks" in Chapter 7, developers must always protect their applications against malicious data introduced through normal user input methods. This can entail both traditional data entry through a forms-based interface and automated data entry through XML Web Services and XML files.
- *Use least privilege.* This principle asserts that by restricting the permissions of the identity under which the code is running, you can limit the amount of damage malicious users can do. Although single-user operating systems such as Windows CE, upon which the Compact Framework runs, do not typically support various permission levels, this principle's application in server software such as ASP.NET can go far toward protecting an application.
- *Obscurity is not security.* This principle has two primary applications. The first is that simply because an application resource (such as a URL when using SQLCE replication) is not well known or published publicly, this does not mean it is not vulnerable to attack. In the case of Web servers, it has been well documented that even anonymously deployed servers record attempts by hackers to break into the systems within hours. The second is that home-grown attempts at security, for example devising custom encryption algorithms, will not

typically be as secure as industry-standard techniques, for example, symmetric encryption algorithms like Triple Data Encryption Standard (Triple-DES).[2]

- *Fail securely.* This principle is a reminder that even if an application is designed with the other principles in mind, an application that gives away its own secrets through default error messages and stack dumps renders adherence to the other principles moot. This is true even in simple cases, for example, an application that simply displays "Invalid Credentials" rather than "Invalid Password," which allows hackers to note when they've discovered a valid user account.

Each of these principles could be discussed in much greater depth. For more information on this topic, we recommend Howard and LeBlanc's *Writing Secure Code* referenced in the "Related Reading" section.

As mentioned in the first principle, security must be applied in depth in order to protect the weakest link. To that end the remainder of this section discusses securing the three principle layers in a Compact Framework application: the device, the application, and the communications layers.

Securing the Device

The first layer of security to consider is that provided by the device itself. Various vendors may employ different mechanisms to secure a device or rely on the facilities of the underlying platform and operating system, such as the Pocket PC running on Windows CE. In addition, organizations can augment the OS software with third-party solutions. Generally, the types of security that can be used to protect the device fall into the categories of authentication, antivirus protection, and lockdown.

Authentication

The simplest step organizations can take to protect devices on which Compact Framework applications run is to require some form of authentication. For example, Pocket PC 2002 and later devices running Windows CE support

[2] Triple-DES is a symmetric encryption algorithm based on the original DES algorithm developed in 1974 and adopted as a standard in 1977. It uses three 64-bit keys (hence the name), rather than one and, therefore, encrypts the data three times.

both simple and strong power-on password authentication for the device.[3] By enabling this feature on the device, the user must type in a simple four-character password or a more complex seven-character password (with upper- and lowercase letters, in addition to numbers or punctuation) at a configurable interval that defaults to one hour as shown in Figure 9-1.

If a user enters an incorrect password, an exponentially increasing time delay is enforced, making it very difficult to guess the password repeatedly. Third-party vendors such as Trust Digital, through its PDASecure product,[4] provide power-on password authentication with additional options that include locking the device after a certain number of invalid password

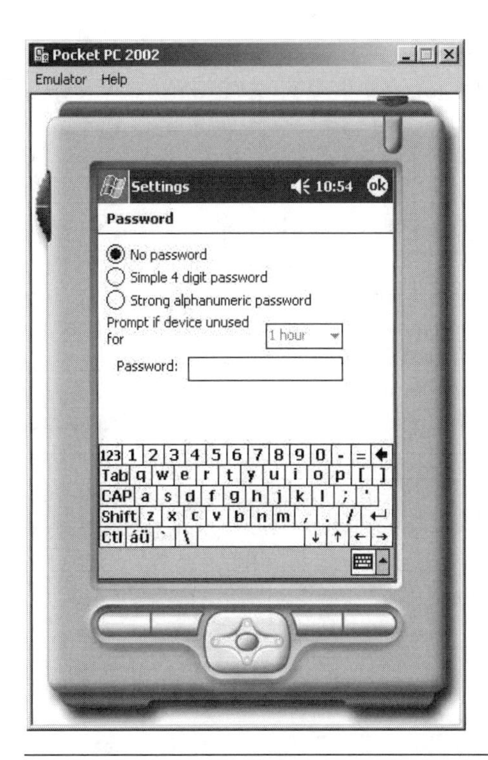

Figure 9–1 *Using Pocket PC Power-On Authentication.* This screen shot shows the interface used to configure the device for power-on password protection. Note that the reauthentication delay is set to one hour by default.

[3] This feature can be downloaded for Pocket PC 2000 by installing Microsoft's Password for Pocket PC at www.microsoft.com/windowsmobile/resources/downloads/pocketpc/powertoys.mspx.
[4] See www.trustdigital.com/prod15b.htm.

attempts and additionally requiring a soft reset, hard reset, or even "wiping" the device of its data.

Although this form of authentication can be quite effective, it is based on knowledge of the password that can perhaps be obtained elsewhere. Implementing authentication based on possession of a physical item, such as a smartcard or a cryptographic certificate located on a removable storage card, or based on who the user is through biometric (fingerprint) or signature identification, requires third-party products. Various vendors such as A2000 Distribution, Certicom Corporation, and Cloakware Corporation provide a variety of such solutions.[5]

Antivirus Protection

Even though viruses that target mobile devices are not as prevalent as those targeting desktop computers, they are still a potential threat to any software on the device. Typically, however, devices are not damaged by viruses but rather pass them into a corporate network via e-mail attachments and documents.

There are, as you might expect, a variety of antivirus packages on the market for devices such as the Pocket PC from vendors that include McAfee, Computer Associates, and F-Secure. In addition, personal firewall products such as VPN-1 are available from vendors such as Check Point Software Technologies.

NOTE: If antivirus and third-party authentication software is installed on the device, it is a best practice to place the software in flash ROM, rather than RAM, on the device. In this way, the software will survive a hard reset (in which the user removes the batteries); however, storing your application in RAM can also work in your favor because the application will be lost during a hard reset, thereby disallowing access to it.

Lockdown

Following the principle of reducing the attack surface, many organizations may wish to disable some features of the device. Third-party software such as PDASecure from Trust Digital, for example, does just that by password protecting or disabling infrared communications, ActiveSync, and the voice recorder.

---

[5] See the white paper "Security on the Pocket PC" referenced in the "Related Reading" section for more information on security options and links to these vendors' products.

Securing the Application

The second layer to which security should be applied is the application itself. Typically, this security includes the concepts of authentication, data protection, and user input.

Authentication

Throughout this book we've highlighted the various ways that a Compact Framework application can connect to back-end data. The primary mechanisms include the following:

- *Invoking SqlClient directly:* This option was discussed in Chapter 4 and involves directly connecting to a remote SQL Server using the SqlClient .NET data provider.
- *SQLCE connectivity:* Both RDA and merge replication use the SQLCE connectivity architecture described in Chapter 7.
- *XML Web Services:* In Chapter 4 we discussed calling XML Web Services from the Compact Framework using the generated proxy class in VS .NET 2003.
- *HTTP Pluggable Protocol:* In Chapter 4 we also discussed how developers can make direct calls to Web sites using the HTTP protocol through the classes in `System.Net` namespace.

In each case, the application may need to provide credentials to the back-end server. In the case where SqlClient is invoked directly, this may simply be the user name and password required by SQL Server or the Windows account credentials, depending on whether Windows authentication is enabled in SQL Server. In the case of SQLCE connectivity, we saw that this may involve sending both the proxy server and IIS authentication credentials, as well as the SQL Server credentials, depending on how both IIS and SQL Server are configured. Finally, in the case of XML Web Services and HTTP, because both techniques make requests to Web servers, the authentication scheme enabled on the particular Web server or proxy server will determine what credentials must be passed.

KEY POINT

In all cases, the best practice, of course, is to prompt the user for the credentials, rather than hard-coding them into the application,[6] and then to

[6] It should be remembered that hard-coding security information in an assembly is not safe because it is easily viewable with utilities such as MSIL Disassembler (ildasm.exe) and classes in the `System.Reflection` namespace.

create the various connection strings or credential objects in memory as necessary. If credentials do, however, need to be stored on the device (for example, if you don't wish to prompt the user for multiple sets of credentials for the proxy, IIS, and SQL Server, or when authenticating through a VPN), they should be stored in an encrypted fashion, in an encrypted SQLCE database (as described in Chapter 5), in a file encrypted using either a third-party solution such as PDASecure or the CryptoAPI, or even on a storage card that the user is required to insert into the device before running the application.[7]

NOTE: Compact Framework developers will note that application configuration files and the associated `Configuration` class in the `System.Configuration` namespace are not supported. In the desktop Framework, developers often use the `Configuration` class to retrieve connection strings and other application settings from the XML configuration file; however, a custom class can easily be created to retrieve such settings using the classes in the `System.Xml` namespace, as discussed in Chapter 3.

Although providing credentials to applications using SqlClient or SQLCE connectivity is fairly straightforward through the use of connection strings and public properties, providing credentials to XML Web Services and directly to Web servers using HTTP requires some custom code.

Using SOAP Headers

In the case of XML Web Services, at the time of this writing, there is no agreed upon industry standard for providing credentials to Web Services, although proposals such as WS-Security from IBM and Microsoft provide the beginnings of standardization in this area (see sidebar titled "WS-Security and More in the WSE"). As a result, most organizations rely on either the authentication provided by the underlying transport (HTTP in most cases) or a custom technique using SOAP headers.

To use custom SOAP headers, the developer of an XML Web Service can simply create a class that inherits from the `SoapHeader` class in the

[7] Although you can store credentials in the Windows CE registry, doing so will only make them harder to find and will not actually protect them (remember, obscurity is not security). Since there are no managed classes in the Compact Framework to manipulate the registry, developers will need to make the appropriate OS calls using PInvoke or download a class library written by another developer (several of which can be found on the Web).

WS-Security and More in the WSE

In fact, Microsoft has released an add-on for both the Windows .NET Framework 1.0 and 1.1 called Web Services Enhancements (WSE) 2.0 (formerly known as the WSDK in its technical preview incarnation). This add-on provides advanced Web Services functionality, including support for credentials, digital signatures, encryption, message routing, and attachments. The functionality of the WSE is based on the WS-Security, WS-Routing, WS-Attachments, and DIME specifications jointly developed by Microsoft and IBM. This add-on includes additional classes and is installed on a developer's workstation to be used when building XML Web Services with VS .NET.

When a WSE-enabled Web Service is consumed in VS .NET, the client application must include a reference to the `Microsoft.Web.Services.dll` assembly. This assembly also includes various dependencies to assemblies and classes that are not included in the Compact Framework (such as `System.Configuration.Install.dll` and many of the classes in the `System.Security.Cryptography` namespace). As a result, the long and short of the matter is that the Compact Framework does not, in its initial release anyway, include the necessary client-side assemblies for consuming WSE-enabled Web Services.

There have, however, been several attempts to use WSE-enabled Web Services from the Compact Framework by employing custom SOAP headers and SOAP extensions and invoking the necessary cryptographic functions of the operating system using the PInvoke service.*

For more information about WSE, see the link in the "Related Reading" section at the end of this chapter.

* See www.learnmobile.net/MobileClient/Tutorials/cfWSE for one attempt at using the beta of the Compact Framework.

`System.Web.Services.Protocols` namespace to expose the credential fields (such as the user account and password) that follow:

```
Public Class AuthHeader : Inherits SoapHeader
    Public UserAccount As String
    Public Password As String
End Class
```

The developer can then allow the client to provide the `SoapHeader` when calling the Web Service by adding a public field to the `WebService` class and decorating the methods that require the header with the `SoapHeaderAttribute`, as the skeleton class shown in Listing 9-1 makes clear.

Listing 9–1 *Implementing a SOAP Header.* This listing shows the skeleton code implemented by a Web Service to require the use of a SOAP header as defined in the previous snippet.

```
<WebService(Namespace:="http://books.atomicdotnet.com ", _
     Name:="AtomicPublishersService")> _
Public Class AtomicPublishers : Inherits
  System.Web.Services.WebService

    Public sHeader As AuthHeader  'SOAPHeader object

    <WebMethod(EnableSession:=False, _
         Description:="Returns orders that need to be fullfilled", _
         BufferResponse:=True), _
         SoapHeader("sHeader", Direction:=SoapHeaderDirection.In)> _
    Public Function GetOrdersToFullFill() As PubOrders
        ' Do a security check
        Me.Authenticate()

        ' Perform the operation and return data
    End Function

    Private Sub Authenticate()

        ' Authorize the use of this Web Service by pulling
        ' the user account and password out of the SOAP Header
        ' as in sHeader.UserAccount and sHeader.Password and
        ' authenticating against a directory.

        ' Throw an exception is not authenticated as in
        ' Throw New SoapException("Incorrect Credentials ", _
        '  SoapException.ClientFaultCode)
    End Sub

End Class
```

You'll notice in Listing 9-1 that the `AtomicPublishers` class includes a public field (`sHeader`) that represents the `SoapHeader` class. The single operation (the `GetOrdersToFullFill` method) then uses the `SoapHeader-Attribute` to identify the object used to process the header, which is sent only from the client to the server. When the client invokes the method, the `sHeader` variable is automatically populated by the .NET Framework, and consequently, its `UserAccount` and `Password` fields can be inspected in the private `Authenticate` method and used to perform the authentication and authorization.

Fortunately for Compact Framework developers, if such an XML Web Service must be called, the proxy class created in the SDP will include the `SoapHeader` class and will allow the credentials to be populated before calling the `GetOrdersToFullFill` method:

```
AtomicPublishersService ws = new AtomicPublishersService();
AuthHeader auth = new AuthHeader();
auth.UserAccount = txtUser.Text;
auth.Password = txtPwd.Text;
ws.AuthHeaderValue = auth;
PubOrders po = ws.GetOrdersToFullFill();
```

Using Web Server Authentication

In a simpler way, classes such as `System.Net.WebRequest` and the Web Service proxy class (inherited from `SoapHttpClientProtocol`) expose a `Credentials` property that can be populated with a `NetworkCredential` object used to provide credentials to a Web server configured with basic or, in the case of IIS, Windows-integrated security.

NOTE: Windows CE, and thus the Compact Framework, does not support Kerberos, Digest, or client certificate authentication.

To add a `NetworkCredential` to a `WebRequest` like that shown in Listing 4-4 where `wreq` is the `WebRequest`, the following code could be written:

```
Dim c As New NetworkCredential("dfoxkc", "notmypassword", "")
wreq.Credentials = c
```

The third argument to the constructor of the `NetworkCredential` object is the domain, which must be supplied if the Web server is using Windows-integrated authentication.

Implementing Time-outs

When a Compact Framework application prompts the user for credentials, remember that (as discussed in Chapter 2) by default the application is minimized and not closed when the X button in the upper right-hand corner of the application is tapped. As a result, a Compact Framework application will remain in memory unless it is explicitly closed by the application code or using the memory applet in the device settings or until a soft reset is performed or the operating system needs to free up memory.

In some cases, this behavior is not ideal because a nontrusted user may be able to pick up a device on which the application is still running and then have complete access to the data and functionality using the already entered credentials.

To avoid this situation, developers can include a class like that shown (with some additional enhancement, to be sure) in Listing 9-2, which tracks the inactivity in one or more forms by using a timer.

Listing 9–2 *Tracking Inactivity.* The class shown in this listing can be used to alert the application to a period of inactivity so that it can reauthenticate the user.

```
Namespace Atomic.CeUtils

    Public Class Inactivity

        Private Shared _t As Timer
        Private Shared _interval As Integer

        Public Shared Event Inactivated(ByVal interval As Integer)

        Public Shared Property Interval() As Integer
            Get
                Return _interval
            End Get
            Set(ByVal Value As Integer)
                _interval = Value
                _t.Enabled = False
                _t.Interval = _interval * 1000
                _t.Enabled = True
            End Set
        End Property

        Shared Sub New()
            _t = New Timer
            _interval = 600 ' 10 minutes by default
            _t.Interval = _interval * 1000
```

```
            AddHandler _t.Tick, AddressOf Inactive
            _t.Enabled = True
        End Sub

        Public Shared Sub RegisterForm(ByVal f As Form)
            ' Hook the events for the control and the form
            Dim c As Control
            For Each c In f.Controls
                AddHandler c.GotFocus, AddressOf WakeUp
            Next
            AddHandler f.Activated, AddressOf WakeUp
        End Sub

        Private Shared Sub WakeUp(ByVal sender As Object, _
         ByVal e As EventArgs)
            ResetTimer()
        End Sub

        Private Shared Sub ResetTimer()
            ' Reset the timer
            _t.Enabled = False
            _t.Enabled = True
        End Sub

        Private Shared Sub Inactive(ByVal sender As Object, _
         ByVal e As EventArgs)
            RaiseEvent Inactivated(_interval)
        End Sub

    End Class

End Namespace
```

In this case, the `Inactivity` class simply exposes a `RegisterForm` static method that accepts a `Form` object and registers handlers for the `GotFocus` events of all of its controls, as well as the `Activated` event for the form itself. When one of these events fires, the timer is reset (the setting defaults to ten minutes, but is configurable through the `Interval` property). In the event that the form (or forms) is inactive for the specified amount of time, the class raises an event, which can then be caught in the application and used to prompt the user for login credentials.

A form in an application can then use the class like so:

```
Atomic.CeUtils.Inactivity.RegisterForm(Me)

AddHandler Atomic.CeUtils.Inactivity.Inactivated, _
  AddressOf ReAuthenticate
```

Data Protection

The second major task in securing an application is to protect the data it uses. As discussed in Chapters 3 and 5, the two major forms of data used by Compact Framework applications are local files (typically in XML format) stored in RAM or on a storage card on the device and SQLCE. In order to secure the data in both cases, encryption should be used.

NOTE: Keep in mind that perhaps the simplest form of data protection is to use the backup and restore facilities provided by ActiveSync. These facilities allow users to perform full and incremental backups and restore the backup onto the device.

SQLCE Encryption

As discussed in Chapter 5, SQLCE allows databases (.sdf files) to be password protected and optionally to be encrypted using the symmetric RC4 algorithm with a key generated from an MD5 hashed password.

Using this approach, all of the data retrieved from a remote SQL Server using RDA or merge replication can be safely stored on the device. An application can then prompt for the password and use it in the connection string to open the database. Be aware of the following issues, however:

- *The password is not recoverable*. The database password cannot be regenerated if lost, so the application will need to be able to recreate the data (for example, through a synchronization) in the event the password is lost and not recoverable.
- *The password can be changed only by compacting the database*. This means that applications will need to include the code required to compact the database as shown in Chapter 5 and that the device will need enough free space to handle the compaction.
- *The strength of the encryption depends on the length of the password*. Because the key the RC4 algorithm uses is a hash of the password,

the length of the password determines the strength of the encryption. The application should enforce at least a minimal (eight-character or longer) password that combines letters, numbers, and special characters.

- *You can create single password applications*. One simple approach is to use SQLCE encryption as the gatekeeper for the entire application. In other words, the application prompts the user for the password and then retrieves the other required credentials (for IIS, SQL Server, and so on) from a settings table within SQLCE. In this way the user needs only a single password, and other credentials are safely stored on the device.

File Encryption

When data is stored in XML files or some other file format on the device or on a storage card, it should be protected via encryption.

As mentioned in Chapter 1, the .NET Framework supports a robust set of encryption classes in the `System.Security.Cryptography` namespace, as discussed in the article referenced in the "Related Reading" section at the end of the chapter. These classes support both symmetric (such as DES, Triple-DES, and RC2) and asymmetric (such as RSA and DSA) algorithms, as well as hashing (MD5, SHA1, SHA512, among others) and random-number generation; however, they are not supported in the Compact Framework. The underlying operating systems, such as Windows CE, may, however, provide system APIs to perform encryption. In fact, Windows CE does include a subset of the CryptoAPI, which is callable using PInvoke.[8]

It is also worth noting that when using the CryptoAPI on Pocket PC 2000, only 40-bit encryption is supported out of the box. In order to use 128-bit encryption, the Microsoft High Encryption Pack can be downloaded.[9] Pocket PC 2002 devices already include this functionality. There are also several code libraries, such as the one offered by Certicom Corporation, that include encryption functionality that can be integrated into an application.

A simpler, and more global, technique might be to use one of the various third-party solutions. For example, Casio provides storage cards with built-in hardware encryption. By storing files or even a SQLCE database on such a card, the application can offload the responsibility for data security.

---

[8] Some of the CryptoAPIs, however, require callbacks, which are not supported in the Compact Framework.

[9] See www.microsoft.com/mobile/pocketpc/downloads/ssl128.asp.

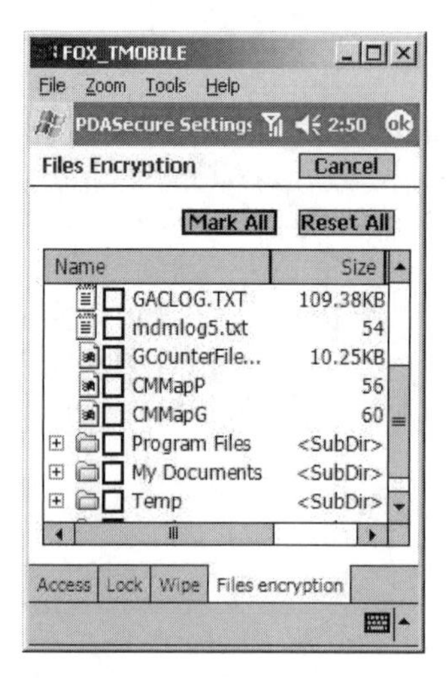

Figure 9–2 *Using File Encryption.* This screen shot shows the interface used by PDASecure to encrypt files stored on a Pocket PC.

In addition, vendors such as Trust Digital, through its PDASecure product, provide file-based encryption (available in a number of different algorithms) tied to the power-on password authentication discussed earlier. For example, the PDASecure application provides a dialog like that shown in Figure 9-2 for encrypting selected files on the device.

Securing User Input

KEY POINT

The final issue to consider when dealing with application security is securing user input as codified in the "Don't trust user input" principle discussed earlier. As discussed in Chapter 7, this issue is most directly relevant when accepting user input that is subsequently used to query or update data in SQLCE or on a remote SQL Server. Malicious users may attempt to insert, or inject, additional SQL into a `TextBox` control, thereby causing a possibly harmful statement to run against the database. These SQL injection attacks can be mitigated by cleansing what the user enters with a function like that shown in the following snippet or using parameterized queries:

```
private string SanitizeIntoSqlLiteral(string strQ)
{
   //Convert single quote into two single quotes
   return strQ.Replace("'", "''");
}
```

In general, user input can be checked using regular expressions through the `RegEx` class of the `System.Text.RegularExpressions` namespace. One interesting use of this class is to create a derived `TextBox` control that uses a regular expression to validate the text entered by the user, as shown in Listing 9-3.

Listing 9–3 *Validating User Input.* This derived `TextBox` control uses the `RegEx` class to validate the input using a regular expression.

```
Public Class PatternTextBox : Inherits TextBox

    Private _r As Regex

    Public Sub New()
        _r = New Regex(String.Empty)
    End Sub

    Public Sub New(ByVal expression As String)
        _r = New Regex(expression)
    End Sub

    Public Property Expression() As String
        Get
            Return _r.ToString
        End Get
        Set(ByVal Value As String)
            _r = New Regex(Value)
        End Set
    End Property

    Private Sub PatternTextBox_Validating(ByVal sender As Object, _
     ByVal e As System.ComponentModel.CancelEventArgs) _
     Handles MyBase.Validating

        If Me.Text = String.Empty Then
            Return
        End If

        If Not _r.IsMatch(Me.Text) Then
            e.Cancel = True
            Me.Select(0, Me.Text.Length)
            Me.Focus()
        End If
```

```
    End Sub
End Class
```

In Listing 9-3 the `PatternTextBox` class inherits from `TextBox` and exposes the `Expression` public property used to hold the regular expression. The `Validating` event is then hooked by the `PatternTextBox_Validating` method and there determines whether the text in the control matches the regular expression. If it doesn't, the text in the control is selected, and focus is not allowed to leave the control.

A client could then create an instance of the `PatternTextBox` control, placing it on the form and ensuring that the user enters a valid address:

```
PatternTextBox p =
 new PatternTextBox("^\w+[\w-\.]*\@\w+((-\w+)|(\w*))\.[a-z]{2,3}$");
this.Controls.Add(p);
```

TIP: Obviously, an interesting enhancement to this control would be to preload it with a variety of regular expressions to handle common types of input, including currency, dates and times, social security numbers, and telephone numbers.

What about CAS?

As many .NET Framework developers are no doubt aware, the .NET Framework includes an enhanced security system called Code Access Security (CAS). CAS provides a security model that is a step beyond the traditional identity-based model (where permissions for running code are derived from the identity of the account under which the code is currently executing) by allowing assemblies to be assigned varying levels of trust based on attributes of the code itself. These attributes can include, but are not limited to, a digital certificate, the originating site or URL, the application directory, the hash value of the assembly, or the strong name.

CAS then includes a series of permissions (defined as classes in the Framework) that are grouped into *permission sets* and then assigned to code based on *code groups,* which define the intersection of permission sets and evidence. Security policies then define the various code groups and permission sets at the user, machine, and enterprise levels. CAS configuration can be set using the

Microsoft .NET Framework Configuration tool installed in the Administrative Tools group. For more information on CAS, see the articles in the "Related Reading" section at the end of this chapter.

Although this model can be very effective for protecting a user's machine and yet allowing an application to execute with the permissions it needs, especially for code in smart client applications downloaded from the Web, the benefits of CAS are not as transferable to smart devices. As a result, the full implementation of CAS in the Compact Framework is not included in the initial release. The Compact Framework does, however, support all of the infrastructure necessary to implement security policy and code groups. It is just that in the initial release, security policy cannot be edited, and all code runs in the `All_Code` code group associated with the `FullTrust` permission set and therefore has full access to the device.

In addition, because smart device operating systems are by nature single user, the Compact Framework also does not support the identity and principal classes (the latter implementing the `IPrincipal` interface) and the associated role-based security mechanisms, as does the .NET Framework. These classes are used to identify the current user and determine whether that user is in a particular role.

Securing Communications

The final aspect to securing an application involves securing the transmission of data across the network. This aspect can involve a wide range of issues including using SSL, VPNs, WEP, and custom SOAP extensions.

Secure Sockets Layer

Perhaps the most fundamental way to secure the communication channel between a device and network is the use of SSL over HTTP. Simply put, SSL uses public- and private-key encryption based on the RSA algorithm, along with the use of a digital certificate to encrypt all of the communication between a client and a server.

Fortunately, support for SSL is built into both Pocket PC 2000 and 2002 devices, although to support the more robust 128-bit keys, the Microsoft High Encryption Pack should be installed on Pocket PC 2000 devices.

Because SSL is used over HTTP, Compact Framework applications can use it when connecting to a back-end server using SQLCE Connectivity,

XML Web Services, or the HTTP Pluggable Protocol, as mentioned earlier in the chapter.

For example, in an application that uses `WebRequest`, SSL can be enabled simply by changing the URL from "http" to "https"; when using SQLCE Connectivity, the `InternetURL` property can include "https" to enable SSL.

NOTE: It should be remembered that SSL does not provide any support for authentication—only for data security through encryption of the communication channel.

On the server side, the IIS server must have a digital certificate installed for the virtual directory in which the resource (the HTML page, for example) resides. The certificate may be one obtained from a trusted CA such as VeriSign or one generated internally within the organization using Microsoft Certificate Server.[10] In the latter case, and as mentioned in Chapter 7, the root certificate for your organization must be installed on the device, as documented in the SQLCE Books Online.

Virtual Private Networks

A second option for encrypting the communication channel is the use of VPNs that rely on protocols such as PPTP and IPSec to both authenticate and encrypt communications.

KEY POINT

While the Pocket PC 2002 and Windows CE .NET 4.1 devices do not support IPSec, they do support PPTP when setting up connections, for example on the Pocket PC 2002, in the Connections tab of the Device Settings dialog. After entering the address of the VPN server for the connection as shown in Figure 9-3, the device will prompt the user for credentials when a connection is made.

As a result, VPNs can be used transparently on top of any type of connection, for example when using the `WebRequest` class or even when using the `Socket` class in the `System.Net.Sockets` namespace.

In order to provide VPN functionality to Pocket PC 2000 devices or to use IPSec, a number of third-party products are available from vendors that include Certicom Corporation and Check Point Software Technologies.

---

[10] For more information on installing certificates, see the FAQs at www.iisfaq.com/default.aspx?View=P20.

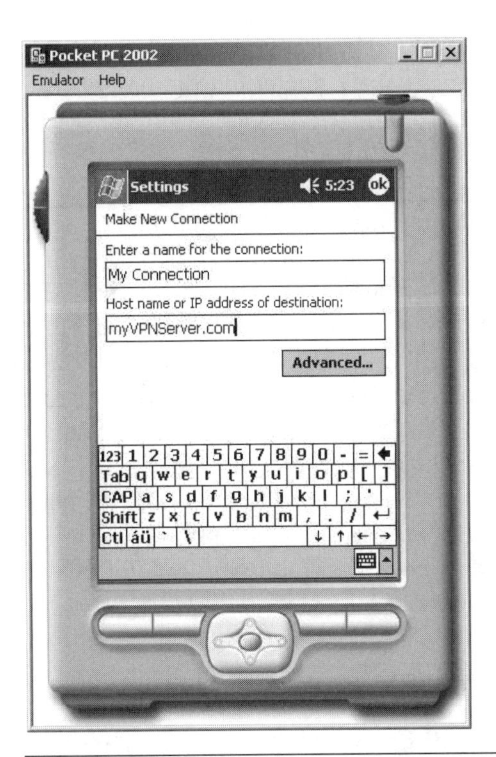

Figure 9–3 *Configuring a VPN*. This screen shot shows the interface for entering the VPN server on a Pocket PC 2002 device.

Wired Equivalent Privacy

An increasing number of devices connect directly to home and corporate LANs through 802.11 protocols. In these networks, unlike in wired LANs, walls are ineffective as a means of security because radio waves are not bound by the walls used to protect the network.

A first-generation technology used to protect such networks was based on the WEP security protocol, which used a shared secret key and the RC4 algorithm; however, early in 2001, a team of researchers at the University of California, Berkeley, discovered flaws in WEP that left it vulnerable to attack, in part because of its use of 40-bit keys and its weak key-management scheme.[11] As a result, WEP is considered nonsecure and should not be used without additional security, such as VPNs, SSL, or patches like that

[11] See the *PC World* article referenced in the "Related Reading" section at the end of this chapter.

created by RSA Security.[12] Although an initiative called WEP2 was developed to address the issues with WEP, including increasing the key size to 128 bits, many in the industry felt that it, too, was vulnerable to attacks.

However, the 802.11i standard, in draft form until the end of 2003, addresses many of these security issues. While the new specification is being ratified, wireless vendors have agreed on an interim solution called Wi-Fi Protected Access (WPA). WPA support is currently being rolled out in products such as Funk Software's Odyssey Client and Meetinghouse Communications's AEGIS client. In the interim other organizations using server and client software primarily from Funk and Meetinghouse have gone forward and implemented the Extensible Authentication Protocol (EAP) over the competing Tunneled Transport Layer Security (TTLS), developed by Funk and Certicom, or Protected Extensible Authentication Protocol (PEAP), developed by Microsoft and Cisco Systems to allow secure access to WLANs on their corporate campuses.

In addition, third parties such as MobileSys and Altarus Corporation offer wireless encryption technology.

Custom SOAP Extensions

Although the other techniques described in this section are more industry standard, developers may implement a custom form of channel security when using XML Web Services by employing a SOAP extension.

Simply put, a SOAP extension is a SOAP pre- and postprocessor that provides extensibility to the normal ASP.NET processing of a Web Service and can be used by both the Web application hosting the Web Service in IIS and in the Compact Framework application that consumes the Web Service on the device. The extension is developed by creating a class that inherits from the `SoapExtension` class in the `System.Web.Services.Protocols` namespace. Within this class, the developer can override methods such as `ProcessMessage` and `ChainStream` to inspect and alter the message, for example, to encrypt and decrypt the body of a SOAP message by calling private methods in the class. The structure of a SOAP extension is shown in Listing 9-4.

[12] See the RSA Wireless Connection newsletter at www.rsasecurity.com/newsletter/wireless/2002_winter/feature.html.

Listing 9–4 *The Structure of a SOAP Extension.* This listing shows how a SOAP extension would be structured. The `ProcessMessage` method is called at various stages during the processing of a SOAP message.

```
public class MyExtension : SoapExtension
{
    Stream oldStream;
    Stream newStream;

    //  Process the message
    public override void ProcessMessage(SoapMessage message)
    {
        switch (message.Stage)
        {
            case SoapMessageStage.BeforeSerialize:
                break;
            case SoapMessageStage.AfterSerialize:
                // User specific code to encrypt the stream
                break;
            case SoapMessageStage.BeforeDeserialize:
                // User specific code to decrypt the stream
                break;
            case SoapMessageStage.AfterDeserialize:
                break;
            default:
                throw new Exception("invalid stage");
        }
    }

    public override Stream ChainStream( Stream stream )
    {
        // Save the stream so the rest of the code can work on it
        oldStream = stream;
        newStream = new MemoryStream();
        return newStream;
    }
}
```

As Listing 9-4 shows, the `MyExtension` class overrides the `ProcessMessage` and `ChainStream` methods. The latter allows the stream to be captured for modification and the former is called at multiple stages during the processing of a SOAP message.

In order for the SOAP extension to be activated, the developer must also create a class inherited from `SoapExtensionAttribute` that is then used to decorate the method(s) for which the extension will be active. For example, for the simple extension shown in Listing 9-4, the following attribute could be created:

```
[AttributeUsage(AttributeTargets.Method)]
public class MyExtensionAttribute : SoapExtensionAttribute
{
    private int priority;

    public override Type ExtensionType
    { get { return typeof(MyExtension); } }

    public override int Priority
    { get { return priority; }
       set { priority = value; } }
}
```

This attribute can then be applied to a method in a Web Service and the corresponding method in the Web Service proxy class. Of course, because the Compact Framework does not support the cryptography classes required to perform the encryption and decryption of the SOAP body, developers will need to build two versions of the extension and attribute classes. In the Compact Framework version, PInvoke calls to the CryptoAPI, or a third-party library can be used, as discussed previously in this chapter. In addition, remember that if a symmetric encryption algorithm is used, both ends of the channel will need to have access to the agreed-upon key to encrypt and decrypt the data. This is not as trivial as it might sound because a nonsecure key exchange leaves the encrypted data vulnerable.

What's Ahead

In this chapter we explained the three primary layers of security that architects and developers need to consider when building Compact Framework solutions. By addressing security at the device, application, and communication layers, the Compact Framework applications you build will be enterprise ready. In the next chapter, we'll continue our discussion of additional programming considerations by considering how to package and deploy Compact Framework applications.

Related Reading

See Microsoft's Trustworthy Computing initiative and related information, at www.microsoft.com/security.

Howard, Michael, and David LeBlanc. *Writing Secure Code*. Microsoft Press, 2002. ISBN 0-7356-1588-8.

Dedo, Douglas. "Security on the Pocket PC," white paper, Microsoft, March 2002, at www.microsoft.com/mobile/enterprise/papers/security.asp.

"Web Services Enhancements (WSE) from Microsoft," at http://msdn.microsoft.com/webservices/building/wse/default.aspx.

Fox, Dan. "Protect Private Data with the Cryptography Namespaces of the desktop Framework." *MSDN Magazine* (June 2002), at http://msdn.microsoft.com/msdnmag/issues/02/06/Crypto/default.aspx.

Fox, Dan. "Secure Your .NET Smart Apps with CAS," at http://builder.com.com/article.jhtml?id=u00220030212dlx01.htm&vf=tt.

Fox, Dan. "Make Managed Code Work with .NET's CAS," at http://builder.com.com/article.jhtml?id=u00220030402dlx01.htm.

"Wireless Security Flawed." *Researchers Report* (February 5, 2001), at www.pcworld.com/news/article/0,aid,40442,00.asp.

CHAPTER 10

Packaging and Deployment

Executive Summary

After a Compact Framework solution has been developed, localized, and secured, it must be packaged and deployed. Fortunately, much of the work involved with packaging an application is taken care of by the SDP functionality in Visual Studio .NET 2003. And, once packaged, the application itself can be deployed in a variety of ways, depending on the connectivity characteristics of the devices. Some of these deployment patterns may also require some custom code development in order to integrate with ActiveSync or to create an auto-updating (a truly smart client) application.

In order to make decisions about how to deploy an application, the versioning system used in the Compact Framework must also be understood. The most important point to keep in mind is that the code-sharing model employed by the Compact Framework, as in desktop Framework, is exactly the opposite of that used by managed environments such as COM. In managed environments such as the Compact Framework, referenced assemblies are copied into the calling application's bin directory and are therefore said to be private assemblies. This allows applications to be isolated from one another and updated independently because two applications are not referencing the same assemblies.

Assemblies may also be explicitly shared by generating a strong name at compile time based on cryptographic information coupled with identity information (assembly name, version, and culture) that uniquely identifies the assembly. The assembly can then be placed in an intelligent registry of sorts on the device called the global assembly cache (GAC). The EE then consults the GAC when making binding decisions for shared assemblies. Manipulating the GAC on a smart device is much different from doing so on the desktop and involves deploying a text file that the EE inspects and uses to update GAC entries. Using the GAC saves space on the device by allowing multiple applications to reference the same shared assembly, and it allows changes to be made in one place, where all applications referencing the assembly can immediately take advantage of it.

To package a Compact Framework application, developers can use the features built into SDP by right-clicking on the project and selecting Build Cab File. SDP then builds .cab files for each supported processor type along with the source .inf and other files needed to customize and rebuild the .cab files. The appropriate .cab file can then be executed on the device, and the application installed.

To deploy the .cab files, various options are available, including using the Application Manager utility provided by ActiveSync, making the .cab files available on a Web site (internal or external) or file share (internal) and using a memory storage card, while taking advantage of the Autorun capability of Pocket PC devices. To ease the deployment of new assemblies in the application, developers can also add an auto-updating capability to Compact Framework applications that allows them to check for new assemblies, using one or more of the distribution methods just enumerated, and then download the updated assemblies before executing the application.

Packaging and Deploying in the Compact Framework

The penultimate chapter both in this book and in this third section of additional programming considerations takes a look at the options for packaging and deploying Compact Framework solutions. As you'll see, much of the work involved with packaging an application is taken care of by the SDP functionality in Visual Studio .NET 2003. Once packaged, the application and the Compact Framework itself can be deployed in a variety of ways, some of which may require custom code development, depending on the location and connectivity of the device.

However, before addressing the two central concerns outlined in the previous paragraph, we'll begin with a discussion of how assemblies are versioned and how the EE handles various versions of the same assembly on the device.

Versioning in the Compact Framework

One of the truly revolutionary aspects of the .NET Framework and, thus, of the Compact Framework is the enhanced support for versioning and code sharing. Unlike previous unmanaged environments such as COM, in which a component was automatically published to the systemwide registry, the managed world supports both private and shared components, as well as a robust side-by-side versioning system.

Private Assemblies

In one key respect, the code-sharing model of the Compact Framework inverts that used by managed environments, such as COM, as exposed through VB 6.0. In the Compact Framework, components (assemblies) that are built and subsequently referenced by other projects are by default private assemblies. In other words, VS .NET by default copies the referenced assembly into the bin directory of the calling project, thereby creating a private copy of the assembly for each application in which it is referenced. This is most easily seen in the Properties window in VS .NET, in which the Copy Local property is set to True for private assemblies.

When the calling project is compiled, the resulting assembly contains a manifest, which includes, among other things, the name of the assemblies that it has referenced, along with their version numbers, referred to as the assembly bindings. The manifest can be viewed using the IL Disassembler utility (ILDasm.exe), which ships with the .NET Framework, as shown in Figure 10-1. At runtime the EE loads a referenced assembly before JIT compiles the first method that refers to it. For private assemblies, the version number is not consulted, and so as long as the referenced assembly exists in the bin directory, it will be loaded and its methods invoked if possible. Of course, if a method in a referenced assembly has its signature changed (by changing the return or parameter types, for example), a runtime error such as `MissingMethodException` will be thrown.

NOTE: The Compact Framework does not support precompiling assemblies to native code through the use of the Native Image Generator (NGen.exe) utility, as the desktop Framework does. This utility is used in the .NET Framework to bypass the JIT compilation and, therefore, to speed the load time of the assembly.

This feature of the Compact Framework, inherited from the desktop Framework, allows great flexibility by allowing developers to update an application simply by deploying an updated assembly in the application's bin directory, an example of which we'll show later in this chapter, and not having to worry about registration. It also promotes application isolation so

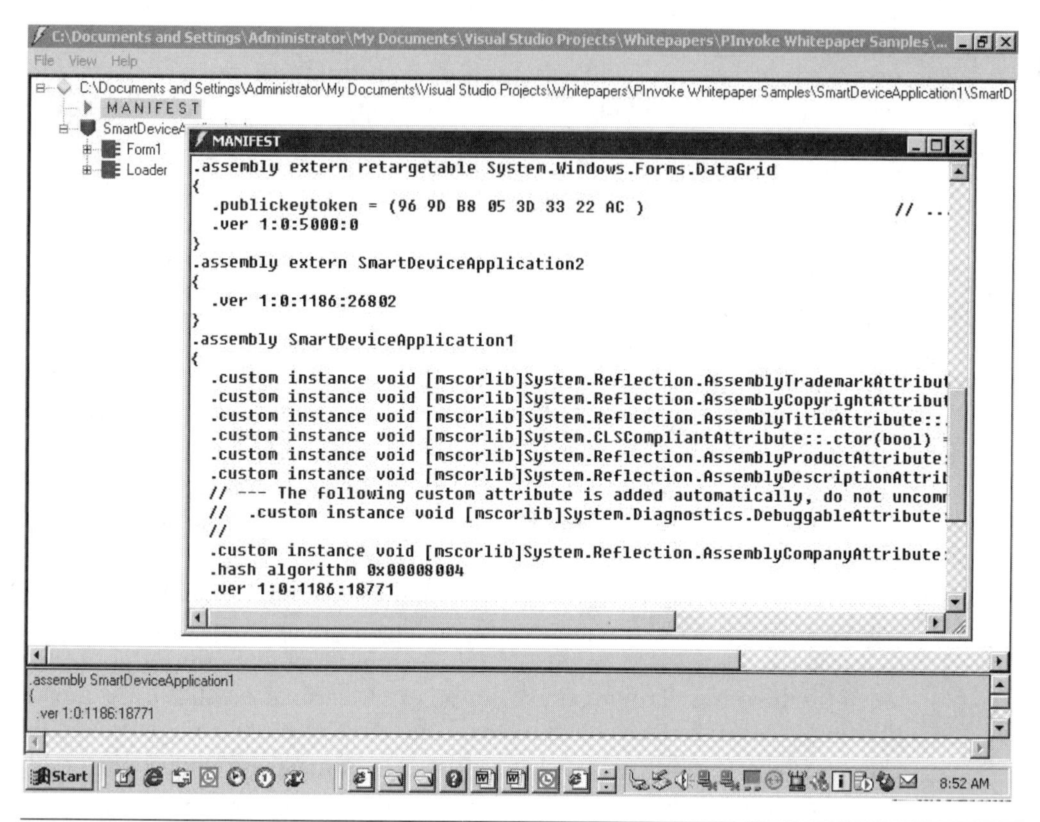

Figure 10–1 *Assembly Binding*. This screen shot shows the manifest of a smart device application using the ILDasm utility. The manifest contains the references to other assemblies. For private assemblies (such as `SmartDeviceApplication2` in this case), the name of the assembly and its version number are recorded.

that changes to one assembly needn't affect other applications installed on the device. This addresses the classic problem of installing a new DLL on a device and inadvertently breaking existing applications.

Shared Assemblies

Although the default behavior of the Compact Framework is to use private assemblies, developers can optionally create shared assemblies that can simultaneously be referenced by multiple applications on the same device. In this way, an updated assembly distributed with a single application can automatically be used by other applications on the device. To create a shared assembly, because sharing is not the default it is in the unmanaged world, a developer must create a strong name at build time and place the assembly in the GAC on the device.

Creating a Strong Name

Simply put, a strong name is information placed into an assembly at compile time that provides name uniqueness, version compatibility, and binary integrity to the shared assembly. It does this by placing a public-key token (called the originator) and a digital signature in the assembly, along with the assembly's identity information (assembly name, version number, and culture or locale information). This information is then used by the EE at binding time to make sure that the correct assembly is loaded and that the assembly has not been altered since it was compiled.[1] In the desktop Framework this verification actually occurs when the assembly is placed in the GAC, instead of at binding or load time.

Because referencing an assembly that has a strong name implies that you want to derive the benefits of strong names, assemblies that use them can reference only other strong-named assemblies. In other words, if you create a class library assembly that relies on code in another assembly, that other assembly must also have a strong name. For this reason, it is a best practice to create strong names for assemblies that contain base-level or common functionality and may be used from a variety of other assemblies. For example, the assemblies that comprise the Compact Framework have strong names that allow them to be referenced from any other assembly, including those with strong names.

To create a strong name for the assembly, just as in the desktop Framework, a developer adds attributes to the AssemblyInfo.cs (or .vb) file. For example, to reference the file that contains the key pair used to generate the public-key token and digital signature, to set the version number, and to identify the culture to be used, the AssemblyInfo.cs file might contain the following attributes:

```
[assembly:AssemblyKeyFile("AtomicUtils.dat")]
[assembly:AssemblyVersion("1.0.*")]
[assembly:AssemblyCulture ("")]
```

[1] The verification processed is performed by using the digitial signature that is embedded in the assembly.

In this case, the attributes point to the AtomicUtils.dat file, specify the four-part version number as 1.0, followed by two autogenerated values, and indicate that the assembly contains culture-neutral resources (the default as discussed in Chapter 8).

The `AssemblyKeyFile` attribute points to a file that contains a key pair generated by the Strong Name (Sn.exe) utility that ships with the .NET Framework. This key pair is assured to be distinct and so provides the name uniqueness guaranteed by shared assemblies. Alternatively, the `AssemblyKeyName` attribute can be used to point to a key pair stored on the developer's machine in the key store provided by the CryptoAPI. Although the desktop Framework also supports the concept of delayed signing through the `AssemblyDelaySign` attribute, the Compact Framework does not.[2]

KEY POINT

The version number is especially important in the Compact Framework because it helps to enforce strict binding for shared assemblies. In other words the four-part number (*`major.minor.build.revision`*) for the referenced assembly must match that found in the calling assembly's manifest. Although this behavior can be overridden in the desktop Framework through binding information placed in an assembly configuration file, as mentioned in previous chapters, the Compact Framework does not support configuration files or applying this type of dynamic binding. As a result, developers will need to recompile the calling assembly if it is to use a new version of a shared assembly.

TIP: Because the `AssemblyVersion` attribute can automatically generate parts of the four-part version number, it is a best practice to specify explicitly the version number for assemblies that are to be placed in the GAC. This gives developers more control over the binding process.

Finally, the `AssemblyCulture` attribute specifies which culture the assembly supports. In fact, this is the attribute that is applied to satellite assemblies that contain resources discussed in Chapter 8. For example, if the attribute contains the German culture "de," the manifest of the assembly would contain the following declaration:

[2] Delayed signing is used by organizations to centralize the process of applying the cryptographic information to the assembly and to protect the private key used in the process. When the `AssemblyDelaySign` attribute is used, the compiler leaves space in the assembly for the signature, which is then applied later and by a centralized group with access to the private key using the Strong Name utility.

```
.assembly MyAssembly
{
  .hash algorithm 0x00008004
  .ver 1:0:1174:22525
  .locale = (64 00 65 00 00 00 )       // d.e...
}
```

By specifying an empty string, the assembly is assumed to contain executable code and culture-neutral resources. As a result, this attribute should be used only for satellite assemblies.

When a shared assembly is referenced by a calling application, its manifest will contain the assembly name, public-key token, version, and culture; therefore, it will ensure that at runtime the caller will bind only to the specific version of the assembly referenced at design time. In addition, by default, the Copy Local property in VS .NET will be set to False when referencing a strong-named assembly.

NOTE: It is important to remember that the EE finds and loads the appropriate assemblies at runtime and, therefore, at design time it doesn't matter where on the developer's workstation a shared assembly is referenced from; however, it is a best practice to group shared assemblies in a folder and deploy that folder to all of the developer's workstations to provide a common way of referencing shared assemblies. Of course, developers could also access read-only copies of the shared assemblies from a source-code control system, such as Visual SourceSafe.

Using the GAC

Because VS .NET sets the Copy Local property to False for shared assemblies, the application needs to be able to find the assembly at runtime. This is the role of the GAC, which acts as a registry for shared components and allows more than one version of the same component to exist side-by-side on the device. At runtime, the EE first searches for shared assemblies in the GAC. This behavior is the same as that for the desktop Framework; however, the plumbing used by the Compact Framework to handle the GAC is quite different.

On the device, the GAC is located in the \Windows directory and simply consists of the assemblies renamed with the prefix "GAC" and appended with the version number and culture, such as GAC_System.Data.Common_v1_0_5000_0_cneutral_1.dll. Assemblies are placed into the GAC by the

Cgacutil.exe utility invoked by the EE each time a Compact Framework application is executed. This utility looks for files with a .gac extension in the \Windows directory that contain lists of shared assemblies to be placed into the GAC.

The .gac file is an ANSI or UTF-8-encoded text file sent to the \Windows directory during the application's deployment, as discussed later in this chapter.[3] A typical .gac file might look as follows:

```
\Program Files\Atomic\AtomicUtils.dll
\Program Files\Atomic\AtomicData.dll
```

Consequently, if the .gac file is deleted from the \Windows directory or is updated, the appropriate changes (insertions and deletions) will be made to the GAC the next time a Compact Framework application is executed.

Benefits

KEY POINT

There are several benefits to using the GAC in your solutions. First, it allows multiple applications to reference the same shared assembly on the device, thereby not taking up the added storage required for private copies on an already resource-constrained device. Second, it allows changes to be made to the shared assembly that all applications referencing the assembly can immediately take advantage of. And, third, the ability to place assemblies in the GAC allows developers to install multiple versions of the assembly on the device and let each application bind to the correct one.

As mentioned previously, the Compact Framework does not support configurable or dynamic version policy, and so, the version of the assembly that an application was compiled against will always be the one that it binds to at runtime.

Packaging Compact Framework Applications

Before an application can be deployed to a device or set of devices, it must be packaged. Fortunately, VS .NET and SDP provide plenty of help in targeting the application to a specific directory and building CAB files for deployment.

---

[3] The Compact Framework installs a .gac file in the \Windows directory that includes the Compact Framework core assemblies.

Setting Project and File Options

Perhaps the most fundamental aspect of deployment that developers need to be aware of before packaging the application is setting the project and file properties in VS .NET.

At the project level, VS .NET supports the Output File Folder property in the Properties window. This property specifies the deployment directory on the device that the project will be installed in. This property is consulted by the Deploy menu options in VS .NET, as well as the CAB file creation that will be discussed in the next section. It defaults to the \Program Files\ *projectname* folder, but this setting can be changed easily.

The second property is the Build Action property shown in the Properties window for each file included in the project. This option can be set to one of four values shown in Table 10-1. One of the build actions shown therein can be associated with each file in an SDP.

Additionally, an icon can be associated with the application in the Project Properties dialog box accessed by right-clicking on the project name in VS .NET and selecting properties.

KEY POINT

Finally, the build configuration can be set from within VS .NET by selecting the Build/Configuration Manager menu. The resulting dialog is shown in Figure 10-2. It is important to remember to change the solution or

Table 10–1 Build Action Settings

Build Action	Purpose
None	Performs no action for the specified file. This can be used when the file is to be included in the project but not compiled or deployed to the device. This is useful for documentation, such as Visio diagrams.
Compile	Compiles the file into the project. This is the setting used for all code files and forms.
Content	Deploys the file along with the project. This is useful for deploying SQLCE databases, XML files, and other supplementary files that the application may use.

Embedded resource	Includes the file in the assembly as a resource. This is useful for adding images and text files to the assembly and later retrieving them. For example, the following line of VB code loads a bitmap (the name is case sensitive) added to the project using this setting:
	
	`Dim a As [Assembly] = [Assembly].GetExecutingAssembly()`
	`Dim b As Bitmap = New Bitmap( _`
	`a.GetManifestResourceStream("logo.gif"))`

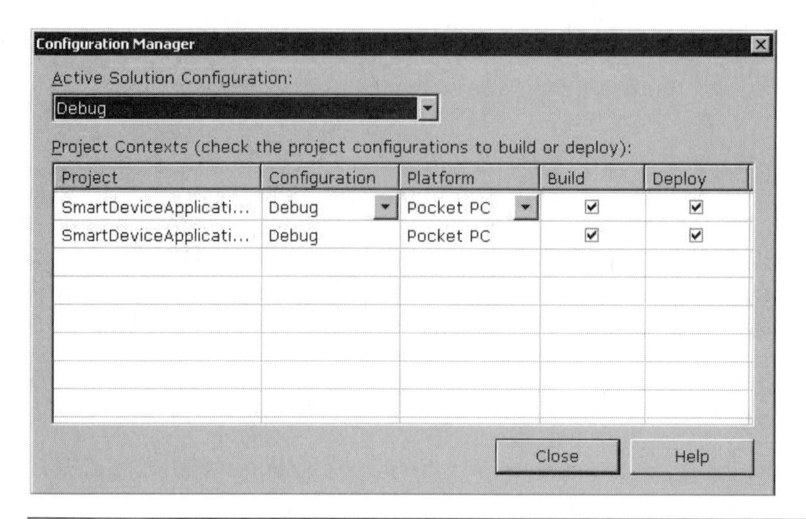

Figure 10–2 *Configuration Manager.* This dialog shows the build and deployment settings for each project in the current solution. Changing from Debug to Release is recommended for performance and size reasons.

project configuration from Debug to Release before deploying the application because the Debug build slows performance and creates larger assemblies. The build mode for the current project can also be changed via the Solution Configuration option in the toolbar in VS .NET.

Creating a CAB File

Once the project options are set, a CAB file can be created to deploy the application. SDP in VS .NET makes this easy by including a Build Cab File menu, accessible by right-clicking the project. Note that this process is different from that in the desktop Framework, where a deployment project that uses the Windows Installer can be added to the solution to package the application.

When this option is invoked, VS .NET silently creates a .cab directory under the project directory on the developer's workstation. In this directory and within a directory identifying the build mode (Debug or Release) are placed a series of .cab files, each targeting a specific processor type, including ARM, ARM4, MIPS, SH3, and X86, and named accordingly, such as MyApplication_PPC_ARM.cab. To install the application on a device, the .cab file simply needs to be copied to the device (as discussed in the third and final section of this chapter) and executed.[4]

[4] WCELoad is the software that ships with Windows CE devices and performs the unpacking and installation of .cab files.

However, in order to customize the installation process, SDP also adds the files needed to customize and rebuild the .cab files in the obj*buildmode* directory. In this directory are placed the BuildCab.bat, Dependencies_*platform*.txt, and *Projectname_platform*.inf files,[5] along with a configuration file for each processor type. The batch file can be used to initiate the rebuild of the .cab files.[6] The dependencies file contains the list of .cab files that this file is dependent on and that always contains a reference to the Compact Framework .cab file for each processor type. At installation time, these dependencies will be checked to ensure that they have been applied to the device. The .inf file contains the installation settings to use when the .cab file is executed.

NOTE: The .cab file created for a project by SDP does not include the .cab file for the Compact Framework. Although future devices will include the Compact Framework in ROM, for the present the Compact Framework .cab file (installed on the developer's workstation) must be copied to the device and executed separately. This is done automatically when the application is deployed from VS .NET using the Deploy menu or when the application is debugged.

To customize the installation on the device, the .inf can be modified. Although a complete description of the .inf file format is beyond the scope of this book, a simple example would be to add a shortcut to the Programs group on a Pocket PC. This can be accomplished by modifying the following sections in the .inf file:

```
[DefaultInstall]
CEShortcuts=Shortcuts

[Shortcuts]
My Application,0,myapp.exe,%CE2%\Start Menu\Programs
```

In this example, the `CEShortcuts` section is redirected to the `Shortcuts` section of the file, and there a shortcut (actually a file with a particular format in Windows CE) is created by specifying (1) the text of the shortcut,

---

[5] Where *platform* is the target platform, for example, PPC for Pocket PC or WCE for Windows CE.

[6] The BuildCab.bat file invokes the CabWiz.exe utility, which, of course, can also be invoked in a stand-alone fashion to customize the build process further. For more information, see the topic "CabWiz.exe for Devices" in the VS .NET 2003 online help.

(2) 0 to identify it as a file, (3) the file to create the shortcut to, and (4) the folder in which to place the shortcut. The identifier `%CE2%` specifies the \Windows directory on a Windows CE device. For more information about the various sections of the .inf file, see the "Related Reading" section at the end of this chapter.

Other examples of customizing the .inf file might include adding additional files (for example, SQLCE databases) to be distributed to the device to the .cab file by modifying the `CopyFiles` section.

Deployment and Installation

After a Compact Framework application has been packaged into a .cab file, it must be deployed and installed on the device. There are several techniques for doing so, including using ActiveSync, a Web site, a storage card, a file share, and even creating an auto-updating application.

Using ActiveSync

Perhaps the most obvious option for deploying a Compact Framework application is to allow the application to be installed when the device is cradled using the ActiveSync software typically used to synchronize files and e-mail, as discussed in Chapter 6.

To do so, the application must be registered with the ActiveSync Application Manager installed on the desktop machine. This process is relatively simple; it requires creating a custom setup executable that invokes the Application Manager on the workstation with ActiveSync installed and passing it an .ini file that contains information about the application to install. For example, to install a simple application called MyApp.exe on the device, the bare-bones .ini file would look as follows[7]:

```
[CEAppManager]
Version      = 1.0
Component    = MyApp

[MyApp]
Description  = My Application
```

[7] The .ini file can also include entries to specify the icon that ActiveSync uses when displaying information about the application.

```
Uninstall     = MyApp
CabFiles      = MyApp_PPC.arm.cab
```

You'll notice that each application to be installed is configured in its own section within the .ini file. In this case, MyApp includes a short description used by the Application Manager when displaying the application in its UI, a pointer to the registry key name used by the Application Manager[8] to uninstall the application if need be, and a comma-delimited list of .cab files to deploy to the device and to install. In this case only the ARM .cab is shown, but others could be added and installed by the Application Manager as necessary.

The custom setup application then needs to invoke the Application Manager and pass it the path to the .ini file. To do so, a developer could write a simple desktop VB .NET application like the one shown in Listing 10-1.

Listing 10–1 *A Custom Setup Application.* This application invokes the Application Manager executable discovered from the registry and passes it the path to the .ini file.

```
Imports System.Diagnostics
Imports System.Reflection
Imports System.IO
Imports Microsoft.Win32

Public Class SetupApp

    Private Const INI_FILE As String = "\setup.ini"

    Public Shared Sub Main()
        Dim ceApp As String = GetCeAppMgr()
        Dim args As String = GetIniArgs()

        ' Start the Application Manager process
        Dim p As Process = Process.Start(ceApp, args)
    End Sub

    Public Shared Function GetCeAppMgr() As String
```

[8] This key defaults to HKEY_LOCAL_MACHINE\Software\Microsoft\Windows\CurrentVersion\Uninstall.

```
        ' Get the key from the registry
        Dim ceApp As String = KeyExists()
        If ceApp = String.Empty Then
            MessageBox.Show( _
         "The Application Manager is not installed on this machine.", _

             "Setup", MessageBoxButtons.OK, MessageBoxIcon.Error)
            Application.Exit()
        Else
            Return ceApp
        End If

    End Function

    Public Shared Function GetIniArgs() As String
        Return """" & Application.StartupPath & INI_FILE & """"
    End Function

    Private Shared Function KeyExists() As String
        Dim key As RegistryKey = Registry.LocalMachine.OpenSubKey( _
      "SOFTWARE\Microsoft\Windows\CurrentVersion\App Paths\
  CEAPPMGR.EXE")

        If key Is Nothing Then
            Return String.Empty
        Else
            Return key.GetValue("", String.Empty)
        End If
    End Function
End Class
```

You'll notice in Listing 10-1 that the program's execution begins in `Sub Main` and first calls `GetCeAppMgr` to retrieve the path to the Application Manager. This path is stored in the registry, and so, if it does not exist, the application throws up a message box and quits. Once the path is found, the Setup.ini file (hard-coded using a constant in this case) is formatted with double quotes and returned through the `GetIniArgs` method. Finally, both the path to the Application Manager and its argument are passed to the shared `Start` method of the `Process` class to start the Application Manager.

The custom setup application can then be packaged (along with the .ini and .cab files) using the Windows Installer in VS .NET 2003 and deployed on the machine that the device connects to. When the user double-clicks the setup application, the application will be deployed to the device.

Although this deployment and installation option is intuitive and provides a familiar mechanism to the user, it does require that the custom setup application be installed on the workstation used for synchronizing with the device.

Using a Web Site

The second option for deploying and installing a Compact Framework application is to use a Web site. Using this option, an organization could create a Internet or intranet site that contains links to the various .cab files created by SDP. This option has the benefit of not requiring the device to be cradled in order to install software.

When a user navigates to the Web page using Pocket Internet Explorer, he or she can tap on the required .cab file. The resulting dialog allows the file to be downloaded and opened. Opening the file after download is equivalent to executing the .cab file on the device and will cause the application to be installed.

NOTE: Regardless of how the .cab file reaches the device, the installation software on the device automatically deletes the .cab file once the installation is complete. To prevent this from happening, the .cab file can be marked as read-only.

To protect the .cab files, especially on a site accessible over the Internet, it is recommend that the virtual directory in which the download page resides be protected by standard HTTP encryption and authentication schemes. For example, the site could use basic authentication to authenticate the user and protect the virtual directory using SSL in order to encrypt both the credentials and the data. Alternatively, a VPN could be used as discussed in Chapter 9.

For scenarios where mobile applications need to be deployed dynamically and are location dependent (an application should be installed on the device when the device enters the WLAN available on a manufacturing floor, for example), third-party provisioning servers can be used. For example, the Appear Provisioning Server (APS) from Appear Network allows automatic delivery, single-click download, and automatic discard of location-specific applications targeted for health care, hospitality, transportation, and workforce coordination.

Using a Storage Card

In many cases, the application needs to be deployed along with a SQLCE database or other software. As a result it can be both time-consuming and bulky to deploy such an application over the Internet or through a cradled connection.

For those reasons, the application can alternatively be deployed on a memory storage card, such as a Compact Flash card; however, rather than requiring the user to execute the .cab file once the storage card is inserted into the device, Pocket PC devices include an Autorun feature.

With this feature, when a storage card is inserted into the device, the Pocket PC looks for an executable called Autorun.exe in a folder mapping to the processor type of the device. For example, if the processor type is ARM, it will then look for the file *Storage Card*\ARM\Autorun.exe on the storage card. When found, the executable is copied to the \Windows folder and executed with the install parameter. Likewise, when the card is removed from the device, the same executable is launched with the uninstall parameter.

As a result, the Autorun executable needs to perform the following steps:

1. *Determine which mode to run in*. Because the application is passed either the install or uninstall command-line parameter, the Autorun application must first determine which of the two modes to run in. The rest of the steps discussed here assume the install mode.[9]
2. *Verify that the application is not already installed on the device.* This can be done by searching for the application's installation directory.
3. *Find the storage card name.* Because storage card names can differ on localized devices (they won't always be called "storage card") and devices can support more than one storage card, techniques such as using the Windows CE `FindFirstFile` API function can be used.[10]
4. *Find the processor-specific .cab file*. This can be done by first detecting the processor on the device and then mapping that to the appropriate directory on the storage card. The Windows CE `GetSystemInfo` returns the processor architecture in its `SYSTEM_INFO` structure.

[9] In uninstall mode, the application will perhaps not be uninstalled, but the Autorun executable could check to see if the application is currently executing and if so display a warning message or shut down gracefully.

[10] See the article "Pocket PC Programming Tips" by John Kennedy at http://msdn.microsoft.com/library/default.asp?url=/library/en-us/dnroad/html/road03132002.asp

5. *Execute the .cab file*. This can be done using a Windows CE API, such as `ShellExecuteEx`.

Because the Compact Framework might not have been previously installed on the device, the Autorun.exe application is typically written using eMbedded Visual C. Further discussion is beyond the scope of this book.

Using a File Share

As with deploying through a Web site, the .cab files can be placed on a network (LAN) share and accessed from the File Explorer on the device, either wirelessly using 802.11 or when cradled using ActiveSync 3.5 and higher.

And, as with deployment through the Web, the share can be protected, and users will be forced to enter valid credentials to gain access to the files. Once again, this option frees the user from having to cradle the device in order to install software.

Setting Up Autodeployment

KEY POINT

Although they differ slightly from the previous deployment options covered in this chapter, Compact Framework applications, like those written with the desktop Framework, can be written to be auto-updating. By auto-updating we mean that the application checks file shares or Web sites either periodically or with each invocation for newer versions of its assemblies and dynamically downloads them. This technique is especially effective for applications that use private assemblies because the application can be partitioned into various functionalities, each housed in a separate assembly and, therefore, updated independently.

The obvious benefit to creating an auto-updating application is that it takes the burden of keeping the application up to date off the user.

To understand how such an application would work, consider the simple `Loader` class shown in Listing 10-2.

Listing 10–2 *An Auto-updating* `Loader`. This class can be used from a Compact Framework application to check automatically for new assemblies on a server and to download them.

```
Public Class Loader

    Protected _assemblies As New ArrayList
    Protected _net As String
```

```
Protected _localPath As String

Public Event Downloading(ByVal f As String)

Public Sub New(ByVal networkPath As String, _
  ByVal assemblies As ArrayList)

    ' Grab the defaults
    _net = networkPath
    _localPath = (Path.GetDirectoryName( _
     [Assembly].GetExecutingAssembly.GetName.CodeBase) & _
     "\").Substring(6)

    ' Load the FileInfo objects
    Dim f As String
    For Each f In assemblies
        If File.Exists(_localPath & f) Then
            Dim fi As New FileInfo(f)
            _assemblies.Add(fi)
        End If
    Next

End Sub

Public Sub CheckForNewAssemblies()

    Try
        Dim fi As FileInfo
        For Each fi In _assemblies
            Dim serverTime As DateTime = _
             Me.GetServerTime(_net & fi.Name)
            If serverTime > fi.LastWriteTime Then
                ' Get the file
                RaiseEvent Downloading(fi.Name)
                Me.GetServerFile( _
                 _net & fi.Name, _localPath & fi.Name)
            End If
        Next
    Catch e As IOException
        ' Exception occurred
        Throw New Exception("Could not retrieve updates")
    End Try

End Sub
```

```
    Protected Overridable Function GetServerTime( _
      ByVal f As String) As DateTime
        ' This method could be rewritten or overridden
        ' to check using HttpWebRequest
        Return File.GetLastWriteTime(f)
    End Function

    Protected Overridable Sub GetServerFile( _
      ByVal serverFile As String, ByVal localFile As String)
        ' This method could be rewritten or overridden
        ' to download using HttpWebRequest
        File.Copy(serverFile, localFile, True)
    End Sub
End Class
```

As Listing 10-2 shows, the `Loader` class exposes a public constructor that accepts the network path of files on the server, along with an `ArrayList` that holds the names of private assemblies to update if new versions are available on the server. The constructor creates an internal `ArrayList` that holds a series of `System.IO.FileInfo` objects to represent each of the private assemblies to inspect.

When the `CheckForNewAssemblies` method is called, it enumerates the assembly list and determines each assembly's date and timestamp on the server, using the `GetServerTime` method. If the server has a newer file, the file is downloaded using the `GetServerFile` method, and it replaces the existing file.

You'll notice that both the `GetServerTime` and `GetServerFile` methods are marked as protected and overridable (virtual in C#). This allows a class derived from `Loader` to override these methods and to use a different means of determining the date and timestamp of the server file and of downloading the file. This could be used to create an `LoaderHttp` class that uses the `HttpWebRequest` class to download new assemblies placed on an intranet.

A bootstrap application or a method in an executable could then use the loader as follows:

```
Dim net As String = "\\myserver\updates\"
Dim s As String = "mydll.dll"
Dim a As New ArrayList
a.Add(s)
```

```
Dim l As New Loader(net, a)
l.CheckForNewAssemblies()
```

What's Ahead

This chapter highlighted the mechanisms built into SDP and VS .NET for packaging and deploying Compact Framework applications. Along the way, we also looked at the issues surrounding the code-sharing model of the Compact Framework and creating shared assemblies. In the final chapter of this book we'll look at some additional developer challenges, many of which are due to the subset nature of the Compact Framework.

Related Reading

Wigley, Andy, and Stephen Wheelwright. *Microsoft .NET Compact Framework, Core Reference*. Microsoft Press, 2003. ISBN 0-7356-1725-2. See especially pp. 207–214 for information on the .inf file structure.

"Successful Installation for Pocket PC Applications," at http://msdn.microsoft.com/library/default.asp?url=/library/en-us/dnppc2k/html/ppc_allation.asp.

CHAPTER 11

Developer Challenges

Executive Summary

Although the Compact Framework, coupled with SDP, provides a rich feature set for building applications, not everything you and your team need is included in the box. Developing professional applications also involves creating applications that perform well. As a result, you'll likely need to augment the Compact Framework and employ some techniques for instrumenting and profiling the code your team writes.

The most obvious way to augment the Compact Framework, which is, after all, a 25% subset of the desktop Framework, is to call unmanaged functions either in the Windows CE API or in custom DLLs that you leverage from existing applications. Some of the reasons to use the Windows CE API include checking the platform of the device, enabling notification, using the RAPI, making phone calls, and processing Short Message Service (SMS) messages, manipulating the registry, and interacting with the system tray.

In order to take advantage of this functionality, Compact Framework developers need to use the PInvoke service. This service (also exposed in the desktop Framework) allows managed code to declare and invoke unmanaged functions residing in DLLs, using the `Declare` statement in VB or the `DllImportAttribute` in both C# and VB. The typical strategy used with PInvoke is to wrap a related set of functions in a class and expose the functionality through static methods.

Although PInvoke in the Compact Framework is very much like that in the desktop Framework, there are subtle differences that include the exclusive use of Unicode and the inability to marshal strings in complex types such as structures. As a result, developers must work around these issues using a variety of low-level techniques.

In addition to augmenting the Compact Framework using PInvoke, Microsoft has made available for download additional tools. These include a remote display control (RDC) for interacting with a cradled device on the desktop, a tool for connecting VS .NET to devices that don't support ActiveSync, the Pocket PC 2003 SDK, which includes new emulators and increased functionality, and a set of handy utilities developed by the Compact Framework team.

Finally, developers always face the challenge of creating applications that perform well. The Compact Framework helps the process by allowing accurate timings to be made for profiling and providing EE performance counters. Developers should also be aware of performance tradeoffs and issues when dealing with data binding, retrieving data, loading forms, and working with strings.

Issues and Challenges

As you've learned throughout this book, the Compact Framework coupled with SDP provides a rich API and a set of features that eases the creation of mobile solutions; however, this doesn't mean that every API and tool that you and your team may need in order to develop your particular solution have been included. In addition, you'll want to consider best practices in order to develop high-performance applications.

In this final chapter we'll look at these sets of issues and concerns by first discussing techniques and tools for augmenting the Compact Framework. Then, we'll explore several coding techniques and their impact on performance.

Augmenting the Compact Framework

To paraphrase Lincoln, you can't please all of the people all of the time, and the Compact Framework is no exception. Because the Compact Framework is a subset of the desktop Framework, covering roughly 25% of its functionality, there will certainly be areas where you and your team will need to implement some functionality that the operating system supports (via the Windows CE API) but that the Compact Framework does not. For this reason, the Compact Framework includes the PInvoke mechanism also common to the desktop Framework. The first parts of this section will therefore provide an overview of PInvoke and its uses and then move on to more advanced issues.

Of course, technology never stands still, and so, on another front, since the release of the Compact Framework in the spring of 2003 (included in

Visual Studio .NET 2003), Microsoft has continued to push forward by providing additional tools and support, including the SDK for the Pocket PC 2003 with devices to be released in the fall of 2003. Because you and your team will want to take advantage of these tools, the second part of this section will provide a rundown of what you should be looking for.

Using PInvoke

In order to take advantage of functionality supported in the operating system, but not in the Compact Framework, developers need to use the PInvoke service. This service (also exposed in the desktop Framework) allows managed code to invoke unmanaged functions residing in DLLs. In this section we'll explore the basics of PInvoke and how its implementation differs from that in the desktop Framework. Then, we'll raise a few of the advanced issues before discussing a couple of common uses of PInvoke for Compact Framework developers.

PInvoke Explained

To use PInvoke to invoke an operating system API, a developer needs to perform three basic functions: declaration, invocation, and error handling.

First, a developer must tell the Compact Framework at design time which unmanaged function is the intended target by specifying the DLL name (also called the module), the function name (also referred to as the entry point), and the calling convention to use. For example, in order to call the `CeRunAppAtEvent` function (used to start a Compact Framework application when ActiveSync synchronization finishes), a developer could use the `DllImportAttribute` or in VB either the `DllImportAttribute` or the `Declare` statement as follows (the C# code snippet is followed by the equivalent in VB):

```
[DllImport("coredll.dll", SetLastError=True)]
private static extern bool CeRunAppAtEvent(string appName,
  ceEvents whichEvent);

Private Declare Function CeRunAppAtEvent Lib "coredll.dll" ( _
  ByVal appName As String, _
  ByVal whichEvent As ceEvents) As Boolean
```

In both cases you'll notice that the declaration includes the name of the DLL in which the function resides (Coredll.dll, in which many of the

Windows CE APIs reside, is analogous to Kernel32.dll and User32.dll in the Win32 API) and the name of the function (`CeRunAppAtEvent`). In addition, the function is marked as `Private` so that it can be called only from within the class in which it is declared. It could also be marked as `Public` or `Friend` (`internal` in C#), depending on how your application will call it.

In the C# declaration, the `static` and `extern` keywords are required to indicate that the function is implemented externally and can be called without creating an instance of the class (although the function is also marked as `private`, so it remains hidden).

You'll also notice that the function requires an argument of type `ceEvents`. In actuality, the function requires a CSIDL value that is a 32-bit integer and maps to one of several Windows CE API constants. In these cases, it is a best practice to expose the constant values (found in the API documentation) in an enumeration, as shown in the following snippet:

```
Public Enum ceEvents
  None = 0
  TimeChange = 1
  SyncEnd = 2
  DeviceChange = 7
  RS232Detected = 9
  RestoreEnd = 10
  Wakeup = 11
  TzChange = 12
End Enum
```

You'll notice that the `SetLastError` property is set to True in the C# declaration (it is False by default). This specifies that the EE calls the Windows CE `GetLastError` function to cache the error value returned so that other functions don't override the value. The developer can then safely retrieve the error using `Marshal.GetLastWin32Error`. In the `Declare` statement, a value of True is assumed, as it is when using `DllImport-Attribute` from VB.

KEY POINT

Second, the developer must invoke the function. As in Windows .NET Framework applications and as discussed in Chapter 3, and as shown in Listing 3-5, it is a best practice to group unmanaged API declarations in a class within an appropriate namespace such as `Atomic.CeApi` and to expose their functions via shared wrapper methods. This class can be packaged in its own assembly and distributed to all of the developers on your team or in your organization. As you might expect, it is a best practice to expose only those arguments in the wrapper that the client needs to supply; the remainder can be defaulted to appropriate values. The class should also contain a private constructor to prevent the creation of instances of the class. A class that abstracts the call to `CeRunAppAtEvent` is shown in Listing 11-1.

Listing 11–1 *A Wrapper Class.* This class wraps the `CeRunAppAtEvent` Windows CE function using a static method.

```
Public Class Environment
   Private Sub New()
   End Sub

   <DllImport("coredll.dll", SetLastError:=True)> _
   Private Shared Function CeRunAppAtEvent(ByVal appName As String, _
    ByVal whichEvent As ceEvents) As Boolean
   End Function

   Public Shared Function ActivateAfterSync() As Boolean
     Dim ret As Boolean
       Try
         Dim app As String
         app = [Assembly].GetExecutingAssembly().GetName().CodeBase
         ret = CeRunAppAtEvent(app, ceEvents.SyncEnd)
         If Not ret Then
           Dim errorNum As Integer = Marshal.GetLastWin32Error()
           HandleCeError(New WinCeException( _
            "CeRunAppAtEvent returned false", errorNum), _
            "ActivateAfterSync")
         End If
         Return ret
       Catch ex As Exception
         HandleCeError(ex, "ActivateAfterSync")
         Return False
       End Try
   End Function

    ' Also create a DeactivateAfterSync method
End Class
```

After the `ActivateAfterSync` method is called, an instance of the application will automatically be started with a specific command-line parameter after the ActiveSync synchronization ends. Additional code is typically added to an SDP in order to detect the command line and activate an existing instance of the application.

Third, developers must be ready to handle two different kinds of errors that can result when an unmanaged function is called. The first is an exception generated by the PInvoke service itself. This occurs if the arguments passed to the method contain invalid data or the function is declared with improper arguments. In this case a `NotSupportedException` will be thrown. Alternatively, PInvoke may throw a `MissingMethodException`, which, as the name implies, is produced if the entry point cannot be found. In the `ActivateAfterSync` method, these exceptions are trapped in a `Try-Catch` block and then passed to another custom method called `HandleCeError`, which can log the exception and perhaps wrap the exception in a custom exception, if needed.

Although the declaration and invocation shown here are the same in the Compact Framework as in the desktop Framework (with the exception of the module name), there are several subtle differences:

- *All Unicode all the time:* In the desktop Framework the default character set, which controls the marshaling behavior of string parameters and the exact entry point name to use (whether PInvoke appends an "A" for ANSI or a "W" for Unicode, depending on the `ExactSpelling` property), can be set using the `Ansi`, `Auto`, or `Unicode` clause in the `Declare` statement and the `CharSet` property in `DllImportAttribute`. Although in the desktop Framework this defaults to ANSI, the Compact Framework supports only Unicode and, consequently, includes only the `CharSet.Unicode` (and `CharSet.Auto`, which equals Unicode) value and does not support any of the clauses of the `Declare` statement. This means that the `ExactSpelling` property is also not supported.
- *One calling convention:* The desktop Framework supports three different calling conventions (which determine issues such as the order arguments are passed to the function and who is responsible for cleaning the stack), which employ the `CallingConvention` enumeration used in the `CallingConvention` property of `DllImportAttribute`. In the Compact Framework, however, only the Winapi value (the default platform convention) is supported, which defaults to the calling convention for C and C++ referred to as `Cdecl`.
- *Unidirectional:* Although parameters can be passed to a DLL function by value or by reference, allowing the DLL function to return data to the Compact Framework application, PInvoke in the Compact Framework does not support callbacks, as the desktop Framework

does, for example, to call the EnumWindows API function to enumerate all of the top-level windows.

- *Different exceptions:* In the desktop Framework, the `EntryPointNotFoundException` or `ExecutionEngineException` will be thrown if the function cannot be located or if it is misdeclared, respectively. In the Compact Framework the `MissingMethodException` and `NotSupportedException` types are thrown.

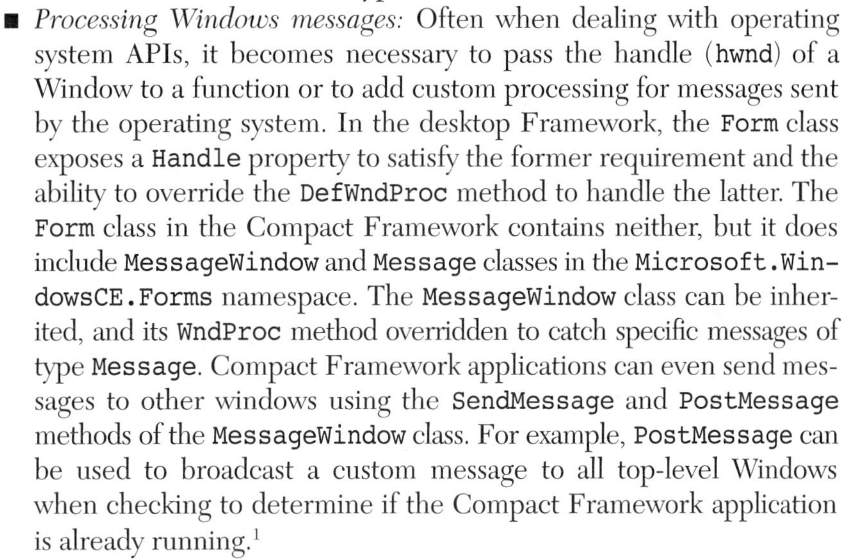

- *Processing Windows messages:* Often when dealing with operating system APIs, it becomes necessary to pass the handle (`hwnd`) of a Window to a function or to add custom processing for messages sent by the operating system. In the desktop Framework, the `Form` class exposes a `Handle` property to satisfy the former requirement and the ability to override the `DefWndProc` method to handle the latter. The `Form` class in the Compact Framework contains neither, but it does include `MessageWindow` and `Message` classes in the `Microsoft.WindowsCE.Forms` namespace. The `MessageWindow` class can be inherited, and its `WndProc` method overridden to catch specific messages of type `Message`. Compact Framework applications can even send messages to other windows using the `SendMessage` and `PostMessage` methods of the `MessageWindow` class. For example, `PostMessage` can be used to broadcast a custom message to all top-level Windows when checking to determine if the Compact Framework application is already running.[1]

Advanced PInvoke

Because the PInvoke service works slightly differently in the Compact Framework than it does in the desktop Framework, there are several issues that developers need to consider when passing data to an unmanaged function.

Passing Strings

A string (`System.String`) in the Compact Framework is a *blittable* type and, therefore, has the same representation as a null-terminated array of Unicode characters in the Compact Framework as it does in unmanaged code.[2] As a result, developers can simply pass strings to unmanaged functions in the same way they would to managed functions, as in the example of `CeRunAppEvent` discussed earlier.

---

[1] See the article referenced in the "Related Reading" section for more information.

[2] Along with `System.Byte`, `SByte`, `Int16`, `Int32`, `Int64`, `UInt16`, `UInt64`, and `IntPtr`.

However, when the string represents a fixed-length character buffer that must be filled by the unmanaged function, as is the case when calling the `SHGetSpecialFolderPath` function shown in Listing 3-5, the string must be initialized before it is passed to the unmanaged function. Because the Compact Framework supports only Unicode, developers can simply declare a new string and pad it with the requisite number of spaces, as shown here:

```
Dim sPath As String = New String(" "c, 255)
```

This behavior is quite different from that in the desktop Framework, where the marshaler must take the character set into consideration. As a result, the desktop Framework does not support passing strings by value or by reference into unmanaged functions and allowing the unmanaged function to modify the contents of the buffer. Instead, both desktop and Compact Framework developers use the `System.Text.StringBuilder` class, as shown in the following snippet:

```
Private Declare Function SHGetSpecialFolderPath Lib "coredll.dll" ( _
    ByVal hwndOwner As Integer, _
    ByVal lpszPath As StringBuilder, _
    ByVal nFolder As ceFolders, _
    ByVal fCreate As Boolean) As Boolean
```

The function can then be called as follows:

```
Dim sPath As New StringBuilder(MAX_PATH)
ret = SHGetSpecialFolderPath(0, sPath, folder, False)
```

Because using the `StringBuilder` is also slightly more efficient, it is recommended when the unmanaged function is passed a string buffer.

Passing Structures

Many Windows CE APIs require that structures be passed to the function. Fortunately, developers can also pass structures to unmanaged functions without worrying, as long as the structure contains blittable types (types represented identically in managed and unmanaged code). For example, the `GlobalMemoryStatus` Windows CE API is passed a pointer to a `MEMORY_STATUS` structure that it populates to return information on the physical and virtual memory of the device. Because the structure contains only blittable types, the structure can easily be called, as shown in the following snippet:

```
Private Structure MEMORY_STATUS
  Public dwLength As UInt32
  Public dwMemoryLoad As UInt32
  Public dwTotalPhys As UInt32
  Public dwAvailPhys As Integer
  Public dwTotalPageFile As UInt32
  Public dwAvailPageFile As UInt32
  Public dwTotalVirtual As UInt32
  Public dwAvailVirtual As UInt32
End Structure

<DllImport("coredll.dll", SetLastError:=True)> _
Private Shared Sub GlobalMemoryStatus(ByRef ms As MEMORY_STATUS)
End Sub
```

In this case, the calling code would instantiate a `MEMORY_STATUS` structure and pass it the `GlobalMemoryStatus` function, as shown here:

```
Dim ms As New MEMORY_STATUS
GlobalMemoryStatus(ms)
```

Because the function requires a pointer to the structure (defined as `LPMEMORYSTATUS` in Windows CE), the declaration of `GlobalMemoryStatus` indicates that the `ms` parameter is `ByRef`. In C# both the declaration and the invocation would require the use of the `ref` keyword.

Alternatively, this function could have been called by declaring `MEMORY_STATUS` as a class, rather than a structure. Because the class is a reference type, the parameter would automatically be passed by reference as a four-byte pointer.

It is also important to note that reference types are always marshaled in the order in which they appear in managed code. This means that the fields will be laid out in memory as expected by the unmanaged function. As a result developers needn't decorate the structure with `StructLayoutAttribute` and `LayoutKind.Sequential`, although they are supported, as in the desktop Framework.

NOTE: Although developers can pass structures (and classes) to unmanaged functions, the Compact Framework marshaler does not support marshaling a pointer to a structure returned from an unmanaged function. In these cases you will need to marshal the structure manually using the `PtrToStructure` method of the `Marshal` class.

Passing Complex Types

KEY POINT

One of the major differences between the marshaler in the Compact Framework and that in the desktop Framework is that the Compact Framework marshaler cannot marshal complex objects (reference types) within structures. This means that if any fields in a structure have types that are not blittable (excluding strings or arrays of strings), the structure cannot be fully marshaled. This is because the Compact Framework does not support the `MarshalAsAttribute` used in the desktop Framework to supply explicit instructions to the marshaler as to how to marshal the data.

In order to solve this problem, Compact Framework developers can employ one of four strategies:[3]

1. *Using a thunking layer:* Simply put, a thunking layer is an unmanaged function that accepts the arguments that make up the structure, that creates the unmanaged structure, and that calls the appropriate function. This is the technique for passing complex objects in structures presented in the Visual Studio .NET help and on MSDN. In order to implement it, however, the developer must create an unmanaged DLL to house the thunking function using eVC++. Developers unfamiliar with eVC++, including eVB developers and desktop Framework developers, will want to use one of the other methods.
2. *Using unsafe string pointers:* The second option for passing strings in structures is to use the `unsafe` and `fixed` keywords in C# (there are no equivalents in VB). While this option allows developers to write only managed code, it does so at the cost of disabling the code-verification feature of the EE, which verifies that the managed code accesses only allocated memory and that all method calls conform to the method signature in both the number and type of arguments. This is the case because using unsafe code implies that developers wish to do direct memory management using pointers.

[3] These three strategies are discussed in detail in our Advanced P/Invoke white paper referenced in the "Related Reading" section at the end of the chapter.

In this approach, the `fixed` keyword is used in conjunction with the `unsafe` keyword and ensures that the GC does not attempt to deallocate an object while it is being accessed by the unmanaged function. In addition to the aforementioned downside of losing code verification, this technique cannot, of course, be used directly from VB. Developers can, however, create an assembly in C# that uses unsafe code and then call that assembly from VB.

3. *Using a managed string pointer:* The third technique used to marshal strings inside of a structure in the Compact Framework is to create a managed string pointer class. The benefits of this approach are that it can be used in both C# and VB and that it can be leveraged in a variety of situations. It does, however, require a little interaction with the unmanaged memory allocation APIs `LocalAlloc`, `LocalFree`, and `LocalReAlloc`.[4]
4. *Custom marshaling:* In some cases, for example when a structure contains fixed-length strings or character arrays, developers may need to resort to custom marshaling because the Compact Framework does not support the `MarshalAs` attribute. This can be done by copying the individual fields of the structure into and out of a byte array, as appropriate, using the methods of the `Marshal` class. A pointer can then be created to the byte array and passed to the unmanaged function.

Common Techniques

There are many reasons that developers might need to use PInvoke to call unmanaged DLLs, a few of which are listed here:[5]

- *Calling custom functionality:* First and foremost, using PInvoke allows your organization to leverage existing code already written and, therefore, to save developer effort. Custom functionality housed in DLLs can then be called using the techniques shown in this chapter; however, there is a slight performance penalty when calling unmanaged code, and, so, developers may see performance improvements by rewriting frequently called unmanaged functions in managed code.

---

[4] An example of this class can be found in our Advanced P/Invoke white paper referenced in the "Related Reading" section at the end of the chapter.

[5] Many of these common techniques have been implemented by developers in the Compact Framework community. Several of the forums developers might look for implementations are listed in the "Related Reading" section at the end of this chapter.

- *Checking the platform of the device:* In order to take advantage of specific functionality on the target platform (Pocket PC versus Windows CE .NET or Pocket PC 2002 versus Pocket PC 2000), developers can use the Windows CE API to check the version and then take the appropriate action (for example, to modify the UI or enable functionality). This can be done by calling the `SystemParametersInfo` API.
- *Enabling notification:* Windows CE includes a notification API (the `CeSetUserNotificationEx` function) that developers can use to display a notify dialog or to cause an application to execute at a specific time or in response to an event, such as at synchronization or when a PC card is changed. Because the Compact Framework doesn't include a managed class that performs this functionality, developers who need it will need PInvoke to make the correct operating system calls.
- *Displaying an icon in the system tray:* Although the desktop Framework includes a toolbox item to allow an application to place icons in the system tray, the Compact Framework does not. As a result, developers can use the `Shell_NotifyIcon` function to place and remove application icons from the system tray.
- *Manipulating the registry:* As mentioned several times in this book, the Compact Framework does not include a managed API to read and write the system registry on the device. Developers can use PInvoke to create a managed wrapper around the requisite Windows CE functions, including `RegOpenKeyExW` and `RegQueryValueExW`.
- *Activating the Pocket PC voice recorder:* Some field service applications require the user to take voice notes using the voice recorder built into some Pocket PC devices. Using an unmanaged function, an application can record the voice note and save it on a storage card.
- *Using the clipboard:* The Compact Framework does not include a class for manipulating the clipboard, and, so, developers will need to wrap the unmanaged `OpenClipboard`, `CloseClipboard`, `GetClipboardData`, and `SetClipboardData` functions in a managed assembly.
- *Using cryptography:* Because the Compact Framework does not include the classes in the `System.Security.Cryptography` namespace, as does the desktop Framework, developers will need to invoke these functions using PInvoke.
- *Making calls on a Pocket PC Phone Edition device:* As mentioned in Chapter 1, Compact Framework applications can be deployed on Pocket PC 2002 (and 2003, as we'll discuss shortly) Phone Edition devices; however, because the Compact Framework does not include a managed API to manipulate the phone functionality, developers

will need to wrap the various functions (exposed in Phone.dll) in their own custom class, such as that shown in Listing 11-2. You'll notice in this listing that the class relies on a custom `StringPtr` class, as discussed in the previous section.

Listing 11–2 *A Phone Class.* This class wraps the functionality needed to make a call on Pocket PC 2002 and 2003 Phone Edition devices.

```
public class Phone
{
  private class PHONEMAKECALLINFO
  {
    private const uint PMCF_DEFAULT              = 0x00000001;
    private const uint PMCF_PROMPTBEFORECALLING = 0x00000002;

    uint cbSize    = (uint)InteropServices.Marshal.SizeOf(
     typeof(PHONEMAKECALLINFO) );
    uint dwFlags            = PMCF_DEFAULT;
    StringPtr pszDestAddress;
    StringPtr pszAppName; // always NULL
    StringPtr pszCalledParty;
    StringPtr pszComment; // always NULL

    public PHONEMAKECALLINFO( string phoneNumber )
    {
      this.pszDestAddress = new StringPtr( phoneNumber );
    }

    public PHONEMAKECALLINFO( string phoneNumber, string calledParty )
    {
      this.pszDestAddress = new StringPtr( phoneNumber );
      this.pszCalledParty = new StringPtr( calledParty );
    }

    public PHONEMAKECALLINFO( string phoneNumber,
     string calledParty, bool prompt )
    {
      this.pszDestAddress = new StringPtr( phoneNumber );
      this.pszCalledParty = new StringPtr( calledParty );
      if( prompt )
        this.dwFlags = PMCF_PROMPTBEFORECALLING;
    }
```

```
        ~PHONEMAKECALLINFO()
        {
          this.pszDestAddress.Free();
          this.pszCalledParty.Free();
        }
      }

      private Phone(){}

      [DllImport ("Phone.dll")]
      private static extern int PhoneMakeCall( PHONEMAKECALLINFO ppmci );

      public static bool MakeCall( string phoneNumber,
       string calledParty, bool prompt )
      {
        PHONEMAKECALLINFO phoneMakeCallInfo = new PHONEMAKECALLINFO(
         phoneNumber, calledParty, prompt );
        return( PhoneMakeCall(phoneMakeCallInfo)==0 );
      }

      public static bool MakeCall( string phoneNumber, string calledParty )
      {
        return( MakeCall(phoneNumber, calledParty, false) );
      }

      public static bool MakeCall( string phoneNumber )
      {
        return( MakeCall(phoneNumber, "", false) );
      }

    }
```

An application can then use the class, as shown here, to make a phone call and prompt the user:

```
Phone.MakeCall("9135551212", "Dan's House", true);
```

Additional Tools

Just as the Compact Framework does not include a managed API for every function that your organization might need, Visual Studio .NET and SDP do not ship with all the tools you'll want to use. Several of the most important tools that you'll want to check out include the following.

Pocket PC Remote Display Control

Even though developers can test their applications using the emulators that ship with SDP in VS .NET 2003, it is always preferred to perform testing on the device itself. To enable a developer to do this easily, Microsoft created the Pocket PC RDC and makes it available in the Power Toys for the Pocket PC on the Microsoft Web site.[6]

This product installs a client (device) and a server (desktop) piece that together allow the device to be manipulated through a window on the desktop. As a result, developers can deploy their applications to the device using VS .NET and then test the application by invoking the RDC and using both the mouse and keyboard for easy user input. The RDC uses TCP/IP and, so, works with ActiveSync connections via Ethernet or dial-up.

TIP: The RDC can be integrated into VS .NET as an external tool by clicking on Tools/External Tools in the IDE, and adding a reference to the C:\Program Files\Remote Display Control\cerhost.exe executable to start the host application on the desktop machine.

Windows CE Utilities for VS .NET 2003 Add-on Pak

A second download from the Microsoft site[7] that you should be aware of is the Windows CE Utilities for VS .NET 2003 Add-on Pak. This add-on addresses the problem of VS .NET 2003 connecting to devices that do not run ActiveSync, when it cannot dynamically determine the CPU on non–Pocket PC devices. As a result, after installing this software, developers should be able to use VS .NET to connect, deploy, and debug applications on any device running Windows CE .NET 4.1 or higher. Specifically, the download includes the following:

- Windows CE CPU Picker for devices with ActiveSync
- Smart Device Authentication Utility for devices without ActiveSync
- Settings to enable debugging with x86-based Windows CE devices
- Components needed to enable debugging with certain devices

---

[6] www.microsoft.com/windowsmobile/resources/downloads/pocketpc/powertoys.mspx.

[7] www.microsoft.com/downloads/details.aspx?FamilyId=7EC99CA6-2095-4086-B0CC-7C6C39B28762&displaylang=en.

SDK for Windows Mobile 2003–based Pocket PCs

Even though this book and the code samples we've presented were developed on the Pocket PC 2002, Microsoft continues to innovate, and in late June of 2003, Microsoft made available the SDK for the Pocket PC 2003 platform. Devices for this platform should be hitting the shelves before this book is printed. The SDK allows developers to use VS .NET 2003 and the Compact Framework to develop applications that target this platform.

Among the changes in the Pocket PC 2003, the most important are that the devices use the Windows CE .NET 4.2 operating system and that the Compact Framework v1.0 is installed in ROM on all Pocket PC 2003 devices. The following is a short list of additional improvements:

- Improved emulators that include drive mapping through folder sharing, that run in three modes (Pocket PC 2003, Pocket PC 2003 Phone Edition, which supports external GSM radio modules, and Pocket PC 2003 Phone Edition with Virtual Radio), and that can be used to test the phone and SMS capabilities of applications
- New and improved BlueTooth support
- Kernel enhancements that lead to improved performance and ROM size savings
- Support for next-generation network layer protocols, such as TCP/IPv6, which supports a significantly larger address space
- Improved display drivers
- New tools, including RapiConfig, an XML-based configuration tool that can be used to configure a device or emulator easily

Additional Developer Tools

In July of 2003, the Compact Framework product team posted to the Microsoft site an additional set of tools that developers can take advantage of, including the following:

- A set of command-line tools that use the RAPI to copy files easily to the device, to debug process on the device, to send debug output to the standard debug window in VS .NET, and to start processes on the device
- A tool called LScript to automate functional testing on the device by capturing and replaying scripts on the device
- A stress-testing tool called Hopper that creates random taps on the emulator to test how the application responds

Measuring and Improving Performance

No matter what the application, architects and developers always must consider the performance implications of their solutions. The Compact Framework is no exception, and, so, in the rest of this chapter, we'll look at the developer challenge of measuring and improving performance in Compact Framework applications.

Specifically, this discussion includes advice on how to measure performance accurately and how to enable performance statistics that are built into the Compact Framework, as well as a discussion of some common performance issues, and finally some general tips on improving performance.

Measuring Performance

In order to assess the design decisions made when implementing an application, an accurate baseline must first be created. The time-honored way of doing this involves implementing a timing mechanism and then instrumenting the code in the application to use the timer.

The most accurate technique available to developers is the use of a high-resolution timer. This mechanism is not provided in the Compact Framework and so requires calling two unmanaged API functions, `QueryPerformanceFrequency` and `QueryPerformanceCounter`. Because the latter function is processor dependent, the former function returns a value that represents the number of counts per second, known as the frequency, while the latter simply returns a count. A developer would call `QueryPerformanceCounter` at the beginning and end of the code block or method, subtract to get the processor-dependent duration, and then divide by the frequency, which would yield a value in seconds. Because some measurements can be short, the code should allow for conversion to milliseconds.

To encapsulate the timing mechanism, a class like `PerfTimer`, shown in Listing 11-3, can be created. This class includes declarations for the two API methods exported from Coredll.dll. In this implementation, the constructor will retrieve the frequency via the `QueryPerformanceFrequency` and two methods for starting and stopping the time measurement. At the beginning of the code block to be measured, the developer would call the `Start` method, which stores the time returned from `QueryPerformanceCounter`. At the end of the code block to be measured, the developer would call the `Stop` method, which returns a 64-bit integer that reflects the number of milliseconds that have elapsed since the `Start` method was called.

Listing 11–3 *A `PerfTimer` Class*. This class wraps the timer functionality needed to instrument a Compact Framework application.

```
using System;
using System.Runtime.InteropServices;

namespace Atomic.CeUtils
{
  class PerfTimer
      {
    [DllImport("coredll.dll")]
    extern static int QueryPerformanceCounter(ref long perfCounter);

    [DllImport("coredll.dll")]
    extern static int QueryPerformanceFrequency(ref long frequency);

    static private Int64 m_frequency;
    private Int64 m_start;

    // Static constructor to initialize frequency.
    static PerfTimer()
    {
      if (QueryPerformanceFrequency(ref m_frequency) == 0)
      {
        throw new ApplicationException();
      }
      // Convert to ms.
      m_frequency /= 1000;
    }

    public void Start()
    {
      if (QueryPerformanceCounter(ref m_start) == 0)
      {
        throw new ApplicationException();
      }
    }

    public Int64 Stop()
    {
      Int64 stop = 0;
      if (QueryPerformanceCounter(ref stop) == 0)
      {
        throw new ApplicationException();
      }
```

```
        return (stop - m_start)/m_frequency;
      }
    }
}
```

After adding this class to a project, it is simple to utilize in an application. There are three steps: (1) create a PerfTimer reference and instantiate, (2) call the Start method at the beginning of the test, and (3) call the Stop method storing the long value at the end of test.

```
Atomic.CeUtils.PerfTimer timer = new Atomic.CeUtils.PerfTimer();
timer.Start();
DoSomething();
long lDur = timer.Stop();
MessageBox.Show("DoSomething executed in " + lDur + "ms");
```

Enabling Performance Statistics

Although using the `PerfTimer` class shown in Listing 11-3 allows an application to profile its own code, it would also be nice to see performance statistics related to the Compact Framework itself, analogous to the Performance Counter statistics available on the desktop through the Performance Monitor utility. Fortunately, this can be accomplished by enabling the Compact Framework performance statistics.

To enable the statistics, a developer need only create the `HKEY_LOCAL_MACHINE\SOFTWARE\Microsoft\.NETCompactFramework\PerfMonitor` registry key on the device. Under this key, a value called `Counters` must be created of type DWORD. To toggle statistics gathering, the value should be set to 1 (on) or 0 (off).

After enabling the statistics, the developer can run a managed program and terminate the process (note that on the Pocket PC, just tapping the X in the upper right-hand corner will not terminate the process), and the report will be generated in a file named Mscoree.stat, located in the root directory on the device.

TIP: It's important not to start any other managed programs while capturing statistics or the statistics may be corrupted.

Table 11–1 Compact Framework Performance Statistics

EE start-up time	Bytes in use after full collection
Total program runtime	Time in full collection
Peak bytes allocated	GC number of application-induced collections
Number of objects allocated	GC latency time
Bytes allocated	Bytes JITed
Number of simple collections	Native bytes JITed
Bytes collected by simple collection	Number of methods JITed
Bytes in use after simple collection	Bytes pitched
Time in simple collect	Number of methods pitched
Number of compact collections	Number of exceptions
Bytes collected by compact collections	Number of calls
Bytes in use after compact collection	Number of virtual calls
Time in compact collect	Number of virtual-call cache hits
Number of full collections	Number of PInvoke calls
Bytes collected by full collection	Total bytes in use after collection

Within the Mscoree.stat file the statistics shown in Table 11-1 will be captured.

Working with the performance counters can be automated by creating an application that can toggle the registry key and display the statistics. Such a utility is discussed in our white paper (along with the meanings of some of the counters listed in Table 11-1) on Compact Framework performance referenced in the "Related Reading" section at the end of this chapter.

Performance Issues

Using the `PerfTimer` and performance statistics allows developers to test the performance of their specific code and its impact on the Compact Framework; however, there are some general performance tests that we've done and placed in our white paper. The most important of these tests and their general conclusions are reproduced here.

Data Binding

In one particular test, both large and small objects in a 100-element array were used to test the performance of automatic and manual data binding to a `ListBox` control. In our tests, manual data binding of the objects in a loop was approximately three times faster than using the `DataSource` property of the `ListBox`. This result should serve to warn developers to test their applications using a variety of techniques, instead of defaulting to the one that is simplest to code.

Handling XML

In a second test, we looked at the differences between using an `XmlTextReader` and an `XmlDocument` object to parse both small and large XML documents on the device. Not surprisingly, our tests confirmed that for simply loading and enumerating a document of any size, the `XmlTextReader` is faster, and as the size of the document increases, the advantage increases as well (from 1.5 times faster to almost 4 times faster). This indicates that unless developers need to hold an XML document in memory, the `XmlTextReader` is the better choice.

XML Web Services and SqlClient

In a third test we looked at the relative performance difference between retrieving data using the SqlClient .NET data provider and using an XML Web Service. Because the Web Service involves creating SOAP messages and using HTTP, whereas SqlClient uses TCP/IP directly, SqlClient outperformed the Web Service by a factor of eight when retrieving data. This indicates that if performance is crucial for applications that are in an always-connected mode, SqlClient is the better choice, unless the database is not a SQL Server.

Loading Controls and Forms

As controls are placed on the designer, VS.NET generates code to instantiate and initialize the controls in the `InitializeComponent` method; however, this code is often not optimized to create the form in a top-down manner by, for example, setting the `Parent` property of containers before adding controls to the container and reducing the number of method and property calls on controls during start-up by using `Controls.Bounds` instead of `Control.Location` and `Control.Size`. In one test, where the form contained a nested hierarchy using panels and a large number of controls, after applying these simple rules, the load time of the form decreased by 50%.

Concatenating Strings

Strings (`System.String`) in managed code are immutable objects, and so, when they are initialized, they retain their value. As a result, if the string is modified through concatenation, it is discarded, and a new string object is created. This has performance consequences when strings are concatenated in loops and during other code-intensive processing. For this reason, a more efficient option is to use the `StringBuilder` class. In our test, a simple concatenation was performed 100 times, and the `StringBuilder` outperformed the string by a factor of 23.

Improving Performance

Finally, the following is a list of more general performance recommendations that developers should keep in mind as they design their applications:

- *Reduce the number of function calls and function size.* Minimize the number of function calls and allow functions to receive more parameters (make the methods "chunkier" and less "chatty").
- *Maximize object reuse.* Frequently, creating objects will lead to performance degradation due to the frequent cleanup of objects and the fragmentation of the managed heap. This is especially important in resource-constrained devices. Keeping objects in memory will reduce the overhead of memory management.
- *Avoid manual garbage collection.* Although the Compact Framework allows applications to invoke the GC, using the `GC.Collect` method, it is recommended that developers stay away from this because most developers will not know better than the Compact Framework when to start the process.
- *Delay initialization.* In order to make an application appear more responsive on start up, the application can delay its initialization by allowing some of it to take place on a background thread. This can be done with the `Thread` class and the `Invoker` class shown in Chapter 3.
- *Use exceptions sparingly.* Throwing exceptions in managed code is a relatively expensive process. As a result, developers should reserve exceptions for truly exceptional cases and not throw them as a normal event in the application.

A Final Word

This chapter ends our exploration of *Building Solutions with the Microsoft .NET Compact Framework*. In this book we've tried to give you a flavor for the design decisions, architecture, and issues and challenges you'll face as you embark on developing great applications with a tool we think is eminently suited for the task. We hope our meager attempt at filling in some of the gaps has been beneficial, and we look forward to hearing your feedback. Good luck!

Related Reading

Fox, Dan, and Jon Box. "An Introduction to PInvoke and Marshaling on the Microsoft .NET Compact Framework," March 2003, at http://smartdevices.microsoftdev.com/Learn/Articles/501.aspx.

Fox, Dan, and Jon Box. "Advanced PInvoke on the Microsoft .NET Compact Framework," March 2003, at http://smartdevices.microsoftdev.com/Learn/Articles/500.aspx.

Fox, Dan, and Jon Box. "Developing Well Performing .NET Compact Framework Applications," July 2003, at http://smartdevices.microsoftdev.com.

Yakhnin, Alex. "Using the Microsoft .NET Compact Framework MessageWindow Class," March 2003, at http://smartdevices.microsoftdev.com/Learn/Articles/515.aspx.

Innovative Decision Support Systems Support Forums, at www.innovativedss.com/forums/default.asp?CAT_ID=4.

.NET Compact Framework Newsgroup, at microsoft.public.dotnet.framework.compactframework.

GotDotNet .NET compact Framework message board, at www.gotdotnet.com/Community/MessageBoard/MessageBoard.aspx?ID=275.

DevBuzz.com .NET Compact Framework Forum.

Index

Wouldn't it be great

if the world's leading technical publishers joined forces to deliver their best tech books in a common digital reference platform?

They have. Introducing **InformIT Online Books** **powered by Safari.**

Specific answers to specific questions.

formIT Online Books' powerful search engine gives you levance-ranked results in a matter of seconds.

Immediate results.

'ith InformIT Online Books, you can select the book ou want and view the chapter or section you need mediately.

Cut, paste and annotate.

iste code to save time and eliminate typographical rors. Make notes on the material you find useful and oose whether or not to share them with your work oup.

Customized for your enterprise.

istomize a library for you, your department or your entire ganization. You only pay for what you need.

---

et your first 14 days FREE!

r a limited time, InformIT Online Books is offering members a 10 book subscription risk-free for days. Visit **http://www.informit.com/online-oks** for details.

informIT Online Books

informit.com/onlinebooks